高等职业教育土建类专业系列教材

土木工程专业英语
SPECIFIC ENGLISH FOR CIVIL ENGINEERING

主　编　帅映勇
副主编　黄　杰
参　编　梁永平

机械工业出版社

本书是为适应我国"一带一路"建设和提高高职高专学生就业质量要求而编写的。本书的主要内容包括国际工程相关规范介绍、FIDIC合同和施工投标、建筑结构、钢筋混凝土、预应力混凝土、岩土工程、测量工程、幕墙、建筑设施、施工过程和技术、施工管理、木建筑、预制装配结构和建筑信息模型，共14个单元和附录。每个单元由学习任务、学习目标、课文、词汇表、注释和课文来源组成。本书课文均选自相关专业英文资料，在课文来源中有介绍；附录介绍了《国际土木工程建筑承包合同》（中英文对照）。全书贯彻现代化、信息化和国际化，努力介绍最新的国际工程动态；内容选材合理、简明易懂、图文并茂，方便读者的学习和理解。

本书可作为高职高专土木工程专业的教材，也可作为土木工程专业的培训教材或参考用书。

为方便教学，本书配有电子课件，凡选用本书作为教材的教师均可登录机械工业出版社教育服务网（www.cmpedu.com）下载，咨询电话：010-88379375。

图书在版编目（CIP）数据

土木工程专业英语/帅映勇主编.
—北京：机械工业出版社，2018.10（2025.1重印）
高等职业教育土建类专业系列教材
ISBN 978-7-111-60806-6

Ⅰ.①土… Ⅱ.①帅… Ⅲ.①土木工程-英语-高等职业教育-教材 Ⅳ.①TU

中国版本图书馆CIP数据核字（2018）第204289号

机械工业出版社（北京市百万庄大街22号　邮政编码100037）
策划编辑：饶雯婧　　　　　　　　责任编辑：饶雯婧
责任校对：徐静姿　张文贵　　　　责任印制：常天培
固安县铭成印刷有限公司印刷
2025年1月第1版·第3次印刷
184mm×260mm·13.75印张·300千字
标准书号：ISBN 978-7-111-60806-6
定价：40.00元

电话服务	网络服务
客服电话：010-88361066	机　工　官　网：www.cmpbook.com
010-88379833	机　工　官　博：weibo.com/cmp1952
010-68326294	金　书　网：www.golden-book.com
封底无防伪标均为盗版	机工教育服务网：www.cmpedu.com

随着国家鼓励企业"走出去""一带一路"和"转型升级，创新发展"等重大战略决策的逐步展开和落实，我国对外承包工程规模日益扩大，合作领域不断拓宽，国际工程市场占有率不断提高，相应的风险也会逐步增加。对外承包工程将不可避免地涉及招投标、规范选用、设备采购、新技术应用等多方面的问题，施工企业迫切需要一批一专多能的国际化土木工程技术应用型人才。

为适应这一发展趋势，编者经过广泛调研与充分研讨，从教材内容入手，采用最新的国际土木工程界规范化语言，开发了实用的土木类英语学习材料，以便更好地培养学生从事涉外土木工程项目所需的基本能力。本书参考了大量的英文专业规范资料，选材合理、内容新颖，能够适应当前的土木工程现状。通过系统的学习，能使读者掌握一定的专业英语词汇并具备一定的英文专业文献阅读能力。

本书由泰州职业技术学院帅映勇主编，梁永平和黄杰参加了前期调研工作、全书的资料提供和内容校核等工作。本书在编写过程中也得到了教研室同事的热忱帮助，在此一并致谢。

由于编者水平有限，书中难免会有缺点和错误，敬请读者批评指正！

编　者

Preface（前　言）

Unit 1　Introduction to Building Code ………………………………………………… 1
Unit 2　FIDIC Contracts and Construction Bids ……………………………………… 13
Unit 3　Building Structure ……………………………………………………………… 28
Unit 4　Reinforced Concrete …………………………………………………………… 40
Unit 5　Prestressed Concrete …………………………………………………………… 51
Unit 6　Geotechnical Engineering ……………………………………………………… 60
Unit 7　Surveying ………………………………………………………………………… 71
Unit 8　Building Facade Types ………………………………………………………… 82
Unit 9　Building Facilities ……………………………………………………………… 92
Unit 10　Construction Process and Technology ……………………………………… 104
Unit 11　Construction Management …………………………………………………… 115
Unit 12　Timber Engineering …………………………………………………………… 126
Unit 13　Prefabricated Building ………………………………………………………… 134
Unit 14　Building Information Modeling ……………………………………………… 141
Appendix　The Contract for Works of Civil Engineering Construction ……………… 151
References（参考文献）………………………………………………………………… 213

Unit 1　Introduction to Building Code

- 学习任务　国际工程相关规范介绍
- 教学时间　2 学时
- 学习目标　通过对美国、欧洲和中国的土木工程规范体系的介绍，基本了解国际工程规范的发展过程，明了国际工程规范的发展现状和趋势，能够适应国际工程建设的有关岗位技术要求。

Text 课文

Building codes have been the primary source for guidance in the design and construction of building structures. For many decades, the application of building codes has been instrumental in safeguarding people's health, safety, and welfare.

1.1　International Building Code

A building code is a set of rules that specify the minimum acceptable level of safety for constructed objects such as buildings. The International Building Code (IBC) is a model building code developed by the International Code Council (ICC). A model building code has no legal status until it is adopted or adapted by government regulation. The IBC provides minimum standards to insure the public safety, health and welfare insofar as they are affected by building construction and to secure safety to life and property from all hazards incident to the occupancy of buildings, structures or premises.[1]

Before the creation of the International Building Code there were several different building codes used, depending on where one decided to construct a building. The IBC was developed to consolidate existing building codes into one uniform code that could be used nationally and internationally to construct buildings. The purpose of the IBC is to protect public health, safety and general welfare as they relate to the construction of buildings.

Therefore, it is used to regulate building construction through use of standards and is a reference for architects and engineers to use when designing buildings or building systems.

1.1.1 History

The first building codes can be traced back to early 1800 BC. The Babylonian emperor Hammurabi enforced what was known as the Code of Hammurabi. This code was very strict and stated that, "If a builder build a house for someone, and does not construct it property, and the house which he built fall in and kill its owner, then that builder shall be put to death." Building codes have evolved over time to protect the safety of building occupants without the threat of death.

Building codes were first seen in the United States in the early 1700's AD. George Washington and Thomas Jefferson encouraged the development of building regulations to provide minimum standards to ensure health and safety of our citizens. In the early 1900's insurance companies lobbied for further development of building codes to reduce property loss payouts caused by inadequate construction standards and improperly built structures. During this time period, local code enforcement officials developed most of the building codes with the assistance of the building industry.

In 1915, the Building Officials and Code Administration (BOCA) was established. This organization developed what is now known as the BOCA National Building Code (BOCA/NBC), which is/was mainly used in the Northeastern United States.

In 1927, the International Conference of Building Officials (ICBO) was established. This organization developed what is now known as the Uniform Building Code (UBC), which is/was mainly used in the Midwest and Western United States.

In 1940, the Southern Building Code Congress International (SBCCI) was founded. This organization developed what is now known as the Standard Building Code (SBC), which is/was mainly used in the Southern United States.

Over the years each of these codes (BOCA/NBC, UBC, & SBC) were revised and updated. Many of the codes were duplications of one another or very similar in nature. In order to avoid duplication and to consolidate the development process BOCA, ICBO, and SBCCI formed the International Code Council (ICC). The purpose of the ICC was to develop codes without regional limitations. In 1994, they began to develop what would become the International Building Code (IBC).

In 1997, the first edition of the IBC was published. There were still many flaws and it was not widely accepted. In 2000, the first comprehensive and coordinated set of the IBC was published. All three organizations (BOCA, ICBO, & SBCCI) agreed to adopt the IBC and cease development of their respective individual codes. The IBC supercedes the

BOCA/NBC, UBC, & SBC codes and states & local governments began to adopt the new consolidated code.

1.1.2 Structure

The International Building Code is arranged in a systematic manner for easy reference. It incorporates all aspects of building construction. It is made up of thirty-five (35) chapters and several appendices. The chapters in the IBC are as follows: 1) Administration; 2) Definitions; 3) Use and Occupancy Classification; 4) Special Detailed Requirements Based on Use and Occupancy; 5) General Building Heights and Areas; 6) Types of Construction; 7) Fire-Resistant-Rated Construction; 8) Interior Finishes; 9) Fire Protection Systems; 10) Means of Egress; 11) Accessibility; 12) Interior Environment; 13) Energy Efficiency; 14) Exterior Walls; 15) Roof Assemblies and Rooftop Structures; 16) Structural Design; 17) Structural Tests and Special Inspections; 18) Soils and Foundations; 19) Concrete; 20) Aluminum; 21) Masonry; 22) Steel; 23) Wood; 24) Glass and Glazing; 25) Gypsum Board and Plaster; 26) Plastic; 27) Electrical; 28) Mechanical Systems; 29) Plumbing Systems; 30) Elevators and Conveying Systems; 31) Special Construction; 32) Encroachments into the Public Right-of-Way; 33) Safeguards During Construction; 34) Existing Structures; 35) Referenced Codes.

Each chapter is broken down into sections and each section into sub-sections. Each section describes performance criteria to be met or references other sections of the IBC or other standards such as ANSI, ASTM, etc. The following is an excerpt from Chapter 7 of IBC 2000.

705.1 General. Each portion of a building separated by one or more fire walls that comply with the provisions of this section shall be considered a separate building. The extent and location of such fire walls shall provide a complete separation. Where a fire wall also separates groups that are required to be separated by a fire barrier wall, the most restrictive requirements of each separation shall apply. Fire walls located on property lines shall also comply with Section 503.2. Such fire walls (party walls) shall be constructed without openings.

As the excerpt indicates, the code structure states regulation in terms of measured performance rather than in rigid specification of materials. This allows for the acceptance of new materials and construction methods without revising the code.

1.1.3 Development

There are five subcommittees of the International Code Council (ICC) that developed and updated the International Building Code (IBC). The Steering and Performance committees of the ICC oversee each of these subcommittees. The committees consisted of code officials (BOCA, ICBO, SBCCI), design professionals, trade professionals, builders and contractors, manufacturers and suppliers, and government agencies.

The development of the IBC typically runs in eighteen-month (18) cycles. The first step is accepting applications for code committees and code change proposals. The next step is to publish the proposed changes. The third step is to hold public hearings on the proposed changes. Next the minutes from the hearing are published. The following step is to collect public comments. The fifth step is to publish the public comments. Next the final public hearing is held. After the final public hearing the annual ICC meeting is held. Finally the revised or new code is published.

New editions of the IBC are published every three (3) years. Amendments to the 2000 edition were issued in 2003 and 2006. In between edition revisions, intervening supplements are published. The last supplement was issued in 2004. The amendments are issued to incorporate approved changes, lessons learned and new technology. All the changes in the new editions are indicated by markings in the margins.

The development of the International Building Code has been an advancement for the building and construction industry. It provides minimum standards to insure the public safety, health and welfare insofar as they are affected by building construction and to secure safety to life and property from all hazards incident to the occupancy of buildings, structures or premises. The IBC is a single source document that is adopted across the United States. This allows contractors to learn one code instead of the several that used to exist depending on the region where the work was performed. Without the IBC or building codes, people would have to think twice before entering structures or their homes.

1.2 EuroCodes

The EN Eurocodes are expected to contribute to the establishment and functioning of the internal market for construction products and engineering services by eliminating the disparities that hinder their free circulation within the community. Further, they are meant to lead to more uniform levels of safety in construction in Europe. The EN Eurocodes are the reference design codes. [2]

The EN Eurocodes apply to structural design of buildings and other civil engineering works including: geotechnical aspects; structural fire design; situations including earthquakes, execution and temporary structures. For the design of special construction works (e.g. nuclear installations, dams, etc) other provisions than those in the EN Eurocodes might be necessary.

The EN Eurocodes cover basis of structural design (EN 1990); actions on structures

(EN 1991); the design of concrete (EN 1992), steel (EN 1993), composite steel and concrete (EN 1994), timber (EN 1995), masonry (EN 1996) and aluminium (EN 1999) structures; together with geotechnical design (EN 1997); and the design, assessment and retrofitting of structures for earthquake resistance (EN 1998).

The Member States of the EU and the European Free Trade Association (EFTA) recognize that EN Eurocodes serve as reference documents for the following purposes: as a means to prove compliance of building and civil engineering works with the basic requirements of the Construction Products Regulation, particularly Basic Requirement 1 "Mechanical resistance and stability" and Basic Requirement 2 "Safety in case of fire"; as a basis for specifying contracts for construction works and related engineering services; as a framework for drawing up harmonised technical specifications for construction products (ENs and ETAs).

1.2.1 Major Concepts of the EN Eurocodes

1. Fundamental requirements (serviceability, safety, fire and robustness)

The structure and structural members should be designed, executed and maintained in such a way that they meet the following:

1) Serviceability requirement—the structure during its intended life, with appropriate degrees of reliability and in an economic way, will remain fit for the use for which it is required;

2) Safety requirement—the structure will sustain all actions and influences likely to occur during execution and use;

3) Fire requirement—the structural resistance shall be adequate for the required period of time;

4) Robustness requirement—the structure will not be damaged by events such as explosion, impact or consequences of human errors, to an extent disproportionate to the original cause.

2. Reliability differentiation

Different levels of reliability may be adopted for both, structural resistance and serviceability. The choice of the levels of reliability for a particular structure should take account of the relevant factors, including: the possible cause and/or mode of attaining a limit state; the possible consequences of failure in terms of risk to life, injury and potential economical losses; public aversion to failure, and social and environmental conditions in a particular location; the expense and procedures necessary to reduce the risk of failure. The levels of reliability that apply to a particular structure may be specified in one or both of the following ways: by classifying the structure as a whole; by classifying its components.

3. Design working life

The design working life is the assumed period for which a structure is to be used for its intended purpose with anticipated maintenance but without major repair being necessary. The notion of design working life is useful for: the selection of design actions (e. g. wind, earthquake); the consideration of material property deterioration (e. g. fatigue, creep); evaluation of the life cycle cost; developing maintenance strategies.

4. Durability

The structure should be designed in such a way that deterioration should not impair the durability and performance of the structure having due regard to the anticipated level of maintenance.

5. Quality assurance

The EN Eurocodes assume that appropriate measures are taken in order to provide a structure, which corresponds to the requirements and to the assumptions made in the design. These measures comprise definition of the reliability requirements, organizational measures and controls at the stages of design, execution, use and maintenance.

1.2.2 EN Eurocode Parts

The EN Eurocodes include 10 standards (EN 1990—1999, as shown in Table 1-1) covering various subjects related to construction.

Table 1-1 Eurocode Parts

	EN Eurocode contents
EN 1990	Eurocode: Basis of structural design
EN 1991	Eurocode 1: Actions on structures
EN 1992	Eurocode 2: Design of concrete structures
EN 1993	Eurocode 3: Design of steel structures
EN 1994	Eurocode 4: Design of composite steel and concrete structures
EN 1995	Eurocode 5: Design of timber structures
EN 1996	Eurocode 6: Design of masonry structures
EN 1997	Eurocode 7: Geotechnical design
EN 1998	Eurocode 8: Design of structures for earthquake resistance
EN 1999	Eurocode 9: Design of aluminium structures

Each of the codes (except EN 1990) is divided into a number of Parts covering specific aspects of the subject. In total there are 58 EN Eurocode parts distributed in the ten

Eurocodes (EN 1990—1999).

All of the EN Eurocodes relating to materials have a Part 1-1 which covers the design of buildings and other civil engineering structures and a Part 1-2 for fire design. The codes for concrete, steel, composite steel and concrete, and timber structures and earthquake resistance have a Part 2 covering design of bridges. These Parts 2 should be used in combination with the appropriate general Parts (Parts 1). Organization of material Eurocodes is shown in Figure 1-1.

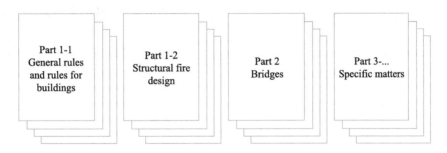

Figure 1-1 Organization of Material Eurocodes

1.3 China Code

China has an extensive system of national and regional building codes and standards which are shown in Table 1-2. The Ministry of Housing and Urban-Rural Development (MHURD) publishes a range of:

- National building codes, generally designated GB or GB/T.
- National sector or industry standards, generally designated JGJ or JGJ/T.

The Municipal Urban-Rural Development and Transportation Commission publishes building codes and standards that address regional requirements. Regional codes are generally designated DBG or DGJ. Enterprise standards (designated 'QB') are published by industry. These are for manufactured products which are not covered by national standards. Provincial and regional jurisdictions enforce mandatory national building codes and standards as minimum requirements. Regional codes cannot set requirements below those required by national codes. Both national and regional codes can include the T-designation which may affect the level of penalties for non-compliance with certain clauses.

China code parts are shown in Table 1-2.

Table 1-2　China Code Parts

Number	English name	Chinese name	Code number
1	Code for Design of Civil Buildings	《民用建筑设计通则》	GB 50352—2005
2	Design Code for Residential Buildings	《住宅设计规范》	GB 50096—2011
3	Residential Building Code	《住宅建筑规范》	GB 50368—2005
4	Code for Investigation of Geotechnical Engineering	《岩土工程勘察规范》（2009年版）	GB 50021—2001
5	Code for Design of Building Foundation	《建筑地基基础设计规范》	GB 50007—2002
6	Load Code for the Design of Building Structures	《建筑结构荷载规范》	GB 50009—2012
7	Code for Design of Masonry Structures	《砌体结构设计规范》	GB 50003—2011
8	Code for Design of Timber Structures	《木结构设计规范》（2005年版）	GB 50005—2003
9	Code for Design of Concrete Structures	《混凝土结构设计规范》（2015年版）	GB 50010—2010
10	Code for Design of Steel Structures	《钢结构设计规范》	GB 50017—2003
11	Technical Code of Cold-Formed Thin-Wall Steel Structures	《冷弯薄壁型钢结构技术规范》	GB 50018—2002
12	Code for Fire Protection Design of Buildings	《建筑设计防火规范》	GB 50016—2014
13	Code for Design of Sprinkler Systems	《自动喷水灭火系统设计规范》	GB 50084—2017
14	Code for Design of Automatic Fire Alarm System	《火灾自动报警系统设计规范》	GB 50116—2013
15	Code for Design of City Gas Engineering	《城镇燃气设计规范》	GB 50028—2006
16	Standard for Lighting Design of Buildings	《建筑照明设计标准》	GB 50034—2013
17	Design code for Heating Ventilation and Air Conditioning	《工业建筑供暖通风与空气调节设计规范》	GB 50019—2015
18	Design Standard for Energy Efficiency of Public Buildings	《公共建筑节能设计标准》	GB 50189—2015
19	Technical Code for Solar Water Heating System of Civil Buildings	《民用建筑太阳能热水系统应用技术规范》	GB 50364—2005
20	Code for Design Protection of Structures against Lightning	《建筑物防雷设计规范》	GB 50057—2010
21	Technical Code for Protection of Building Electronic Information System against Lightning	《建筑物电子信息系统防雷技术规范》	GB 50343—2012

（续）

Number	English name	Chinese name	Code number
22	Assessment Standard for Green Building	《绿色建筑评价标准》	GB/T 50378—2014
23	Technical Code for Glass Curtain Wall Engineering	《玻璃幕墙工程技术规范》	JGJ 102—2013
24	Technical Specification for Application of Architectural Glass	《建筑玻璃应用技术规程》	JGJ 113—2015
25	Technical Code for Metal and Stone Curtain Walls Engineering	《金属与石材幕墙工程技术规范》	JGJ 133—2001
26	Technical Specification for Retaining and Protection of Building Foundation Excavations	《建筑基坑支护技术规程》	JGJ 120—2012
27	General Technical Specification for Mechanical Splicing of Bars	《钢筋机械连接通用技术规程》	JGJ 107—2003
28	Code for Acceptance of Construction Quality of Building Foundation	《建筑地基基础工程施工质量验收规范》	GB 50202—2002
29	Code for Acceptance of Constructional Quality of Masonry Structures	《砌体结构工程施工质量验收规范》	GB 50203—2011
30	Code for Acceptance Construction Quality of Timber Structures	《木结构工程施工质量验收规范》	GB 50206—2012
31	Code for Acceptance of Construction Quality of Building Ground	《建筑地面工程施工质量验收规范》	GB 50209—2010
32	Code for Construction Quality Acceptance of Building Decoration	《建筑装饰装修工程质量验收规范》	GB 50210—2001
33	Code for Acceptance of Construction Quality of Water Supply Drainage and Heating Works	《建筑给水排水及采暖工程施工质量验收规范》	GB 50242—2002
34	Code of Acceptance for Construction Quality of Ventilation and Air Conditioning Works	《通风与空调工程施工质量验收规范》	GB 50243—2016
35	Code for Acceptance of Energy Efficient Building Construction	《建筑节能工程施工质量验收规范》	GB 50411—2007
36	Unified Standard for Constructional Quality Acceptance of Building Engineering	《建筑工程施工质量验收统一标准》	GB 50300—2013
37	Code of Acceptance of Construction Quality of Building Electrical Engineering	《建筑电气工程施工质量验收规范》	GB 50303—2015
38	Code for Acceptance of Installation Quality of Lifts Escalators and Passenger Conveyors	《电梯工程施工质量验收规范》	GB 50310—2002

(续)

Number	English name	Chinese name	Code number
39	Code for Installation and Commissioning of Sprinkler Systems	《自动喷水灭火系统施工及验收规范》	GB 50261—2017
40	Code for Installation and Acceptance of Fire Alarm System	《火灾自动报警系统施工及验收规范》	GB 50166—2007
41	Evaluating Standard for Excellent Quality of Building Engineering	《建筑工程施工质量评价标准》	GB/T 50375—2016
42	Code for Design of Fire Protection for Fossil Fuel Power Plants and Substations	《火力发电厂与变电所设计防火规范》	GB 50229—2006

China's rapid economic development over the past two decades has dramatically changed its position in the world economy. China International Contractors Association announced the official publication of the first batch of English version of the Chinese engineering and technical standards on November 29, 2013, and agreed to use "CHINACODE" as official logo for foreign marketing and distribution, which marks the substantive step of the Chinese standard's "going out", and lays the foundation for next foreign marketing of the Chinese standard in more areas (Figure 1-2).[3]

中国工程技术标准对外推广官方标志

Figure 1-2　China Code Logo

In the context of fierce competition in the international engineering market, the choice of technology standard has become an important means of international competition. The technology standard of a country not only embodies the advanced nature of the country's technology industry, but also has important influence on the discourse power and profit point of the country's enterprises in international projects.[4]

Glossary 词汇表

building code 建筑规范
building construction 建筑施工
consolidated [kənˈsɒlɪdeɪtɪd] adj. 加固的，整理过的，统一的，团结的
fire-resistant adj. 耐火的，防火的
interior finish 室内装饰

criteria [kraɪˈtɪəriə] n.（批评、判断等的）标准，准则
means of egress 疏散设施
accessibility [əkˌsesəˈbɪləti] n. 可达性，易接近，可到达
interior environment 室内环境

energy efficiency 能源效率
exterior wall 外墙
roof assembly 屋顶组件
rooftop structure 屋顶结构
structural design 结构设计
soil and foundation 地基与基础
concrete ['kɒŋkri:t] n. 混凝土
aluminum [ælja'mɪnɪəm] n. 铝
masonry ['meɪsənri] n.
 石工工程，砖瓦工工程，砖石建筑
steel [sti:l] n. 钢，钢铁
wood [wʊd] n. 木材，木制品
glazing ['gleɪzɪŋ] n.
 上釉，玻璃窗，玻璃装配
gypsum board 石膏板
plaster ['plɑ:stə(r)] n. 石膏，灰泥
plastic ['plæstɪk] n. 塑料制品
electrical [ɪ'lektrɪkl] adj. 与电有关的
mechanical system 机电系统，机械系统

plumbing ['plʌmɪŋ] n. 管道工程
elevator and conveying system
 电梯和输送系统
encroachment [ɪn'krəʊtʃmənt] n.
 侵入，侵占，侵蚀
circulation [ˌsɜ:kjə'leɪʃn] n. 传播，发行
timber ['tɪmbə(r)] n. 木材，木料
serviceability [sɜ:vɪsə'bɪlɪti] n.
 有用性，适用性，可维护性
deterioration [dɪˌtɪəriə'reɪʃn] n.
 恶化，退化，变坏
durability [ˌdjʊərə'bɪləti] n.
 耐久性，持久性
substantive [səb'stæntɪv] adj.
 真实的，大量的，实质的
fierce [fɪəs] adj.
 激烈的，凶猛的，狂热的
discourse ['dɪskɔ:s] n.
 论述，交谈，正式的讨论

Notes 注释

[1] A building code is a set of rules that specify the minimum acceptable level of safety for constructed objects such as buildings. The International Building Code (IBC) is a model building code developed by the International Code Council (ICC). A model building code has no legal status until it is adopted or adapted by government regulation. The IBC provides minimum standards to insure the public safety, health and welfare insofar as they are affected by building construction and to secure safety to life and property from all hazards incident to the occupancy of buildings, structures or premises.

  建筑规范是一组规则，规定建筑物等建筑物体的最低可接受安全等级。国际建筑规范（IBC）是国际规范委员会（ICC）制定的示范建筑规范。示范建筑规范在政府法规采纳或修改之前，无法律地位。IBC 提供最低标准，以确保公共安全、健康和福利，因为它们受建筑施工的影响，并在所居住的建筑物、结构或房产发生灾害事件时确保生命和财产的安全。

[2] The EN Eurocodes are expected to contribute to the establishment and functioning of

the internal market for construction products and engineering services by eliminating the disparities that hinder their free circulation within the community. Further, they are meant to lead to more uniform levels of safety in construction in Europe. The EN Eurocodes are the reference design codes.

EN Eurocodes 预计将有助于建筑产品和工程服务内部市场的建立和运作，消除阻碍区域内部自由流通的差距。此外，它们旨在促使欧洲的建筑安全等级更加统一。EN Eurocodes 是参考设计规范。

[3] China's rapid economic development over the past two decades has dramatically changed its position in the world economy. China International Contractors Association announced the official publication of the first batch of English version of the Chinese engineering and technical standards on November 29, 2013, and agreed to use "CHINACODE" as official logo for foreign marketing and distribution, which marks the substantive step of the Chinese standard's "going out", and lays the foundation for next foreign marketing of the Chinese standard in more areas.

过去 20 年来中国快速的经济发展极大改变了它的世界经济地位。中国对外承包工程商会 2013 年 11 月 29 日宣布正式出版发行首批英文版中国工程技术标准，并统一使用"CHINACODE"官方标志对外推广发行，这标志着中国标准"走出去"迈进了实质性的一步，也为下一步更多领域的中国标准对外推广奠定了基础。

[4] In the context of fierce competition in the international engineering market, the choice of technology standard has become an important means of international competition. The technology standard of a country not only embodies the advanced nature of the country's technology industry, but also has important influence on the discourse power and profit point of Chinese enterprises in international projects.

在当前国际工程市场竞争激烈的背景下，技术标准的选择已成为国际竞争的重要手段。采用哪个国家的技术标准，不仅是该国行业技术先进性的体现，也对该国企业在国际项目中获取话语权及利润点有着重要影响。

References 课文来源

[1] The International Construction Magazine http://cwmags.com/
[2] The Global Voice of Consulting Engineers http://fidic.org/
[3] BSI Shop-Buy British Standards https://shop.bsigroup.com/
[4] ICC | International Code Council https://www.iccsafe.org/
[5] Eurocodes: Building the Future http://eurocodes.jrc.ec.europa.eu/home.php
[6] 国家工程建设标准化信息网 http://www.risn.org.cn/Default.aspx
[7] 中国对外承包工程商会 http://www.chinca.org/

Unit 2 FIDIC Contracts and Construction Bids

> - 学习任务　FIDIC 合同和施工投标
> - 教学时间　2 学时
> - 学习目标　初步了解国际咨询工程师联合会各类合同的相关知识及其国际工程招标和投标业务相关的术语，着重对国际工程招投标的具体流程进行仔细介绍，为学生在今后的相关岗位职能提供知识储备。

Text 课文

The International Federation of Consulting Engineers (commonly known as FIDIC, acronym for its French name Fédération Internationale Des Ingénieurs-Conseils) is an international standards organization for the consulting engineering & construction best known for the FIDIC family of contract templates.

Its members are national associations of consulting engineers. Founded in 1913, FIDIC is charged with promoting and implementing the consulting engineering industry's strategic goals on behalf of its Member Associations and to disseminate information and resources of interest to its members. The founding member countries of the FIDIC were Belgium, France and Switzerland. Today, FIDIC membership covers 102 countries of the world.

FIDIC, in the furtherance of its goals, publishes international standard forms of contracts for works and for clients, consultants, sub-consultants, joint ventures and representatives, together with related materials such as standard pre-qualification forms.[1]

FIDIC also publishes business practice documents such as policy statements, position papers, guidelines, training manuals and training resource kits in the areas of management systems (quality management, risk management, business integrity management, environment management, sustainability) and business processes (consultant selection, quality based selection, tendering, procurement, insurance, liability, technology transfer, capacity building). FIDIC organizes the annual FIDIC International Infrastructure Conference and an extensive programme of seminars, capacity building workshops and training courses.[2]

2.1 An Overview of the FIDIC Forms of Contract

2.1.1 FIDIC contract forms

FIDIC was written in a "user friendly" and simple language, with clear and logical structure. Moreover, the FIDIC forms of contracts are consistent in their language and structure with each other making it easy and practical to set up two, or even more contracts for the same work (e.g., employer-contractor and contractor-subcontractor) with minimal conflicts and adjustments between the contracts. At the same time, each of the contracts is complete and can stand by itself. Most importantly however, the FIDIC standard forms have been tested by the industry for extensive period of time in many jurisdictions all over the world. This not only allowed the FIDIC forms to improve, but also allowed the construction industry to get used to, and like, the FIDIC forms. [3]

Over the years FIDIC has consistently improved on its contracts and it has become the tradition that FIDIC contracts are known in popular parlance by the colour of their cover. The organization has added new forms of contract, replaced previous ones and updated important terms. The Table 2-1 gives a brief overview of FIDIC contracts to date.

Table 2-1 FIDIC Contract Forms

FIDIC contract	Year released	Notes
The (old) Red Book (The Construction Contract)	First published in 1957, the fourth and final edition was published in 1987, with a supplement added in 1996	These contracts were aimed at the civil engineering sector, as differentiated from the mechanical/electrical engineering sector
The (old) Yellow Book (The Plant and Design-Build Contract)	First published in 1967 with the third and last edition in 1987	These contracts were aimed at the mechanical/electrical engineering sector
The Orange Book (Design-Build and Turnkey)	The first and only edition of this contract was released in 1995	This was the first design and build contract released by FIDIC
The (new) Red Book (The Construction Contract)	Released in 1999	The Red Book is suitable for contracts that the majority of design rests with the Employer

(续)

FIDIC contract	Year released	Notes
The (new) Yellow Book (The Plant and Design-Build Contract)	Released in 1999	The Yellow Book is suitable for contracts that the contractor has the majority of the design responsibility
The Silver Book (The EPC/Turnkey Contract)	Released in 1999	The Silver Book is for turnkey projects. This contract places significant risks on the contractor. The contractor is also responsible for the majority of the design
The Pink Book (The MDB Construction Contract)	First published 2005—an amended version was published 2006, with a further edition in June 2010	This is an adaptation of The Red Book created to fit the purposes of Multilateral Development Banks
The Gold Book (Conditions of Contract for Design, Build and Operate Projects)	Released in 2008	This is FIDIC's first Design-build and operate contract

Other contracts in the FIDIC family include the FIDIC sub-contract, The Blue-Green Book (Dredgers Contract), which is concerned with dredging and reclamation works, The White Book (Client/Consultant Model Services Agreement), which is for the engagement of consultants by Employers, and The Green Book (The Short Form), which is used mostly for simple, repetitive, short duration jobs.

2.1.2 General features of FIDIC contracts

Although the FIDIC family covers a wide range of contracts, there are some common features:

1. Presentation

FIDIC is usually divided in two parts: Part I consisting of the general conditions and Part II concerning the conditions of particular application (including guidelines for the preparation of Part II clauses). The general conditions are the standard combination of contract provisions, while the particular conditions are the result of negotiation between the parties and are designed to modify or delete some of the general conditions.

Part I contains the general terms of the contract, such issues as rights and obligations of each party, procedure for payment, variation, certification and dispute resolution.

Part II of the contract is the conditions of particular application and is to be used to introduce project specific clauses, such as language of the contract, choice of law, the name of the person or firm appointed to act as Engineer or Employers representative for the project among other terms. The Appendix usually contains sample of documents to be used for the procurement process.[4]

In most FIDIC forms there is a default hierarchy for the documents forming the contract. The order of priority is as stated below and in the event of inconsistency the first on the list takes precedence:

—The Contract Agreement;

—The Letter of Acceptance (this is the formal acceptance of the contractor's tender and marks the formation of the contract);

—The Letter of Tender;

—the conditions of particular application (Part II);

—general conditions of contract (Part I);

—The Specification and Drawings (Red Book), The Employer's Requirements (Yellow Book), the Schedules (Red and Yellow Books);

—Further documents (if any), listed in the Contract Agreement or in the Letter of Acceptance.[5]

2. Dispute resolution

FIDIC contracts adopt a multi-tier dispute resolution process. The emphasis in recent years has been on the amicable settlement of disputes. The process usually provides as a first step, for disputes to be submitted for adjudication before an Engineer or a Dispute Board. If one (or both) of the parties is dissatisfied, a period is allowed for amicable settlement. If the parties are not able to settle the dispute during the "amicable settlement" period, the final stage is to proceed to arbitration. FIDIC contracts provide as a default position that the arbitration rules of the International Chambers of Commerce should apply in the arbitration of disputes arising from the contract.[6]

3. Bias for English law

The first sets of FIDIC contracts were based on English law principles. This bias was so strong, that in commenting on the FIDIC Red Book, first edition, Ian Duncan Wallace QC put it lightly thus: "As a general comment, it is difficult to escape the conclusion that at least one primary object in preparing the present international contract was to depart as little as humanly possible from the English conditions."

Since 1957, future FIDIC contracts have successfully incorporated the principles of other legal systems especially the civil law system. However, the basic framework of

English law principles has survived. For instance, provisions relating to liquidated damages have been maintained.

2.2 Construction Bids

Construction Bids are written offers from contractors to undertake a construction job in return of a certain sum of money. The job can be one or more of the following:
—Construction work defined by drawings and specifications;
—Supply of requisite materials of specified quality;
—Supply of labor required to complete a specified work;
—Transport of materials.

2.2.1 Invitation to tender

When all the preliminaries are completed and the owner has decided to proceed with the work, tenders are invited. Legally this is an attempt to check if there would be interested contractors to carry out the work within the estimated limit of time and finance. The invitation to Tender is not binding to the owner to proceed with the work and does not cause any liability for any expenses to which contractors would spend in preparing and submitting their construction bids.

2.2.2 Bids

Bids can be either:
—Negotiated Bid;
—Limited competition or selective Bid;
—Open competition Bid.

1. Negotiated Construction Bid

In this method the price to be paid in return of the work to be done is negotiated with a single contractor. This, obviously, does not provide the owner with a comparative price. Though the cost of a work will be higher in this method, an owner may expect some advantages from employing a particular contractor whose policies and methods are known and who has in the past proved capable of fulfilling his obligations. The higher cost may be offset by better quality, early completion, and smooth administration. [7]

The negotiated construction bid procedure can be adopted to the owner's advantage if the chosen contractor is one in which the Owner and the Engineer have confidence, and which is of known integrity and reliability. Moreover, the work to be carried out is within his special scope and experience.

The procedure of the negotiated bid is as follows:

—The Engineer on behalf of the owner, invites a contractor to submit a bid. The initial invitation includes information regarding the proposed contract procedure, a brief description of the work, the approximate dates of commencement and completion of the works as well as other essential information;

—The Engineer or the contractor prepares the priced bill of quantities. In many cases it would be more practical that the contractor does the original pricing, as he is in a better position to judge the correct price, which may depend on construction equipment and methods of execution to be adopted by him;

—The priced bills are then handed over to the other party for consideration;

—The rates are examined by the other party. Rates in dispute are compared with current rates for similar work, obtained in competitive tendering after allowing for the special features of the situation;

—When the Engineer and the Contractor reach agreement, the agreed schedule of prices is sent to the Owner for his approval and assent.

2. Limited Competition Construction Bids

In this method the Owner calls for bids from a few contractors who are known to have specialized in the special type of work and from whom he has, in the past, experienced excellent results. This procedure is usually adopted for private works, where the owner has the right to negotiate directly and to enter into agreement with whomsoever he chooses. This procedure is recommended for the public sector when the work requires specialized knowledge such as industrial construction.[8]

3. Open Competition Construction Bids

In this method bids are called by public advertisement. The method is usually adopted for public works, as the rules generally require that government and other public contracts be advertised publicly, to obtain the most advantageous terms. Any contractor who is willing to undertake a work and who has the requisite finance and construction equipment to complete it satisfactorily is allowed to submit a bid.[9]

This open competition procedure has the following advantages:

—It provides equal and just opportunities to all contractors;

—It protects the government from possible insinuations of favoritism;

—It opens the way to having the work done at a minimum cost to the public.

The drawback is the possibility of having a lowest bid submitted by contractor who is unsuitable for carrying out the work and thereby resulting in waste of time and effort.

2.2.3　Information to be given in a Call for Bids Notice

The notice must be as short as possible, but conveying an adequate idea of the nature and scope of the proposed work and all essential details.

The text of a good advertisement should at least include the following information:

—Mode of submitted bids: Bidders should be asked to submit bids in sealed covers, in order to maintain secrecy of quotations.

—Form of bid: It is advisable to get all offers in the same form to facilitate scrutiny, and comparison.

—Name of the inviting authority: This helps the bidder know the persons he will have to deal with and the co-operation he may receive, if his bid is accepted, and such may also affect his quotation.

—Nature of the work and its location: If the nature and location of work is within his operating area, the prospective contractor will want to learn more about the job; else he may not waste time in reading further details.

—Estimated cost of the work: This most briefly indicates the magnitude of the work and enables him to see whether it is too small to interest him or perhaps too large for him to handle.

—Time limit: The time limit within which the work is to be completed should be mentioned. This will be of great help to the bidder to work out a realistic price of the work. The time limit may influence the type and number of construction equipment and workers to be employed, and will vitally affect the contractor's bid.

—The availability of data and forms: where, from whom, at what cost and up to which date blank bid forms and specifications may be obtained and whether a refund will be made upon their return in satisfactory conditions.

—Earnest money required with the bid: The amount and form of security; whether the amount will be accepted in the form of cash, bank guarantee, check or others. Also the number of days within which the amount will be refunded to the unsuccessful bidder.

—Performance security: The amount, form, and procedure of recovering.

—Information regarding drawings: where and when drawings can be examined by bidders. Last date, place and time of receipt of sealed bid.

—The date, time and place and procedure of opening bids.[10]

However, the lowest construction bid is likely to be rejected on account of any of the following cases among others:

—Improper offer: e.g. an offer to execute the work at a cost of so much dollars below the lowest offer is not a proper offer;

—Inadequate finance: The lowest bidder may not have sufficient finance to handle the job;

—Lack of experience in similar works;

—Inadequate construction equipment and staff.

2.2.4 Tips for consideration in preparing construction bids

1. Careful study of the contract documents

The contract documents issued to the contractor or supplied for his inspection include general contract conditions, drawings, specifications, bills of quantities etc. These documents are to be studied carefully to check if any unusual conditions, specifications, or any feature of the work would demand special attention during pricing.

2. Subcontractor's work

Make relative inquiries about prices with sub-contractors and material suppliers for their respective portions of the job.

3. Site visit

This is important to ascertain the conditions under which the work has to be carried out and their impact on pricing.

—Difficulty to access the site;

—Limited space on the site for vehicle movements;

—Type of soil and depth of water table;

—Availability of space for storing materials on site;

—Availability of materials, their sources and prevailing market prices;

—Local availability of skilled and unskilled labor, prevailing wages for workmen;

—Source and cost of water needed for construction;

—Power and lighting source, and the cost of erecting, marinating and dismantling power connection to the site.

4. Time for completion

The bidder then estimates the length of time the work will take and the number and category of permanent staff suggested by the nature of the work for construction management. This helps to calculate the establishment charges.

5. Temporary works

The value of any temporary works needed to commence the construction and to clear away on completion, such as temporary office required for construction management purpose, store sheds for building materials, access road, water supply, depreciation of construction equipment, insurance, taxes, etc.

Glossary 词汇表

Federation of Consulting Engineers 国际咨询工程师联合会
disseminate [dɪˈsemɪneɪt] vt. 散布，传播
pre-qualification n. 资格预审
quality management 质量管理
risk management 风险管理
business integrity management 企业诚信管理
environment management 环境管理
sustainability [səˌsteɪnəˈbɪləti] n. 持续性，能维持性，永续性
consultant [kənˈsʌltənt] n. （受人咨询的）顾问，咨询者
procurement [prəˈkjuːmənt] n. 采购，获得，取得
insurance [ɪnˈʃʊərəns] n. 保险费，保险
liability [ˌlaɪəˈbɪləti] n. 责任，债务
technology transfer 技术转让
capacity [kəˈpæsəti] n. 容量，性能，生产能力
parlance [ˈpɑːləns] n. 腔调，说法，用语
release [rɪˈliːs] vt. 释放，发布，发行，放开
Multilateral Development Bank（MDB）多边开发银行
dredging [ˈdredʒɪŋ] n. 疏浚
reclamation [ˌrekləˈmeɪʃn] n. 开垦，回填
engagement [ɪnˈɡeɪdʒmənt] n. 契约，协议
employer [ɪmˈplɔɪə(r)] n. 雇主
Red Book 红皮书（施工合同条件）
Yellow Book 黄皮书（设备与设计—建造合同条件）
Orange Book 橘皮书（设计及建造合同条件）
Silver Book 银皮书（设计—采购—施工EPC/交钥匙合同条件）
Pink Book 粉皮书（施工合同条件多边开发银行MDB协调版）
Gold Book 金皮书（设计—建造—运营合同条件）
Blue Book 蓝皮书（疏浚与吹填工程合同条件）
White Book 白皮书（业主咨询工程师标准服务协议书）
Green Book 绿皮书（简明合同格式）
tender [ˈtendə(r)] n. 投标
bid [ˈbɪd] n. 投标，（尤指拍卖中的）出价
contract [ˈkɒntrækt] n. 合同，契约，协议
dispute [dɪˈspjuːt] n. 争端，辩论，纠纷
bias [ˈbaɪəs] n. 倾向
provision [prəˈvɪʒn] n. 规定，条项，条款
liquidated damages 规定的违约偿金，清偿损失额
contractor [kɒntˈræktə] n. （建筑、监造中的）承包人（商）
specification [ˌspesɪfɪˈkeɪʃn] n. 规格，规范，明细单，说明书
evaluation [ɪˌvæljuˈeɪʃn] n. 评估
award [əˈwɔːd] vt. 授予，获得，判定 n. 奖品，（法院）判决
negotiate [nɪˈɡəʊʃieɪt] vi.& vt. 谈判，协商，交涉
competition [ˌkɒmpəˈtɪʃn] n. 竞争，比赛，竞争者
earnest money 定金
subcontractor [ˌsʌbkənˈtræktə(r)] n. 转包商
depreciation [dɪˌpriːʃiˈeɪʃn] n. （资产等）折旧，货币贬值，跌价

Notes 注释

[1] FIDIC, in the furtherance of its goals, publishes international standard forms of contracts for works and for clients, consultants, sub-consultants, joint ventures and representatives, together with related materials such as standard pre-qualification forms.

FIDIC 为促进其目标,为客户、咨询师、副咨询师、合资企业和代表出版了符合国际标准格式的工程合同,以及提供资格预审标准格式等相关材料。

[2] FIDIC also publishes business practice documents such as policy statements, position papers, guidelines, training manuals and training resource kits in the areas of management systems (quality management, risk management, business integrity management, environment management, sustainability) and business processes (consultant selection, quality based selection, tendering, procurement, insurance, liability, technology transfer, capacity building). FIDIC organizes the annual FIDIC International Infrastructure Conference and an extensive programme of seminars, capacity building workshops and training courses.

FIDIC 还发布了管理系统领域(质量管理、风险管理、企业诚信管理、环境管理和可持续发展)和业务流程领域(顾问选择、质量选择、招标、采购、保险、责任、技术转让和能力建设)的政策声明、立场文件、指导方针、培训手册和培训资源包等商业实践文件。FIDIC 组织了年度 FIDIC 国际基础设施会议和广泛的研讨会计划、能力建设讲习班和培训课程。

[3] FIDIC was written in a 'user friendly' and simple language, with clear and logical structure. Moreover, the FIDIC forms of contracts are consistent in their language and structure with each other making it easy and practical to set up two, or even more contracts for the same work (e.g., employer-contractor and contractor-subcontractor) with minimal conflicts and adjustments between the contracts. At the same time, each of the contracts is complete and can stand by itself. Most importantly however, the FIDIC standard forms have been tested by the industry for extensive period of time in many jurisdictions all over the world. This not only allowed the FIDIC forms to improve, but also allowed the construction industry to get used to, and like, the FIDIC forms.

FIDIC 采用"用户方便"和简单的语言编写,结构清晰合理。此外,FIDIC 合同格式在它们的语言和结构上是相互一致的,使得为同一项工作(例如雇主—承包商和承包商

—分包商）设立两份甚至更多合同变得简单而实用。这些合同之间的冲突和调整最小，同时，每一份合同都是完整的，可以独立存在。但最重要的是，FIDIC 标准格式已在全世界许多司法管辖区进行了长时间的应用。这不仅允许改进 FIDIC 格式，而且还允许建筑业熟悉 FIDIC 格式。

[4] FIDIC is usually divided in two parts: Part Ⅰ consisting of the general conditions and Part Ⅱ concerning the conditions of particular application (including guidelines for the preparation of Part Ⅱ clauses). The general conditions are the standard combination of contract provisions, while the particular conditions are the result of negotiation between the parties and are designed to modify or delete some of the general conditions.

Part Ⅰ contains the general terms of the contract, such issues as rights and obligations of each party, procedure for payment, variation, certification and dispute resolution.

Part Ⅱ of the contract is the conditions of particular application and is to be used to introduce project specific clauses, such as language of the contract, choice of law, the name of the person or firm appointed to act as Engineer or Employers representative for the project among other terms. The Appendix usually contains sample of documents to be used for the procurement process.

FIDIC 通常分为两部分：第一部分由通用条款组成，第二部分关于专用条款（包括准备第二部分条款的指导原则）。通用条款是合同条款的标准组合，而专用条款是双方当事人之间协商的结果，旨在修改或删除某些通用条款。

第一部分包含合同的通用条款，如当事人的权利义务、付款程序、变更、认证和争议解决等问题。

合同的第二部分是专用条款，用于引入项目特定条款，如合同的语言、法律的选择、被指定担任该项目的工程师或雇主代表的人或公司的名称。附录通常包含用于采购过程的文件示例。

[5] In most FIDIC forms there is a default hierarchy for the documents forming the contract. The order of priority is as stated below and in the event of inconsistency the first on the list takes precedence:

—The Contract Agreement;
—The Letter of Acceptance (this is the formal acceptance of the contractor's tender and marks the formation of the contract);
—The Letter of Tender;
—The conditions of particular application (Part Ⅱ);
—General conditions of contract (Part Ⅰ);

—The Specification and Drawings (Red Book), The Employer's Requirements (Yellow Book), the Schedules (Red and Yellow Books);

—Further documents (if any), listed in the Contract Agreement or in the Letter of Acceptance.

在大多数FIDIC格式中，形成合同的文档有一个默认层次结构。优先级顺序如下所述，如果不一致，列表中的第一个优先：

——合同协议书；

——中标通知书（这是对承包商投标的正式接受，标志着合同的订立）；

——投标书；

——专用条款（第二部分）；

——通用条款（第一部分）；

——技术规范和图纸（红皮书）、雇主的要求（黄皮书）、资料表（红皮书和黄皮书）；

——进一步文件（如有），在合同协议书或中标通知书中列出。

[6] FIDIC contracts adopt a multi-tier dispute resolution process. The emphasis in recent years has been on the amicable settlement of disputes. The process usually provides as a first step, for disputes to be submitted for adjudication before an Engineer or a Dispute Board. If one (or both) of the parties is dissatisfied, a period is allowed for amicable settlement. If the parties are not able to settle the dispute during the 'amicable settlement' period, the final stage is to proceed to arbitration. FIDIC contracts provide as a default position that the arbitration rules of the International Chambers of Commerce should apply in the arbitration of disputes arising from the contract.

FIDIC合同采用多层次争议解决程序，近年来的重点是在争议和解上。该流程通常的第一步是将争议提交至工程师或争议委员会裁决。如果一方（或双方）不满意，则后续允许存在一个和解期。如果双方在"和解期"期间无法解决争议，最后一步是进行仲裁。FIDIC合同默认规定，合同产生的争议仲裁应采用国际商会的仲裁规则。

[7] **Negotiated Construction Bid**

In this method the price to be paid in return of the work to be done is negotiated with a single contractor. This, obviously, does not provide the owner with a comparative price. Though the cost of a work will be higher in this method, an owner may expect some advantages from employing a particular contractor whose policies and methods are known and who has in the past proved capable of fulfilling his obligations. The higher cost may be offset by better quality, early completion, and smooth administration.

议标

在这种方法中，为完成的工作所需支付的价格是与单个承包商协商的。显然，这并不能为业主提供可比较的价格。虽然这种方法的工作成本会更高，但业主可能会期望使用一个特定的承包商可以获得一些优势，这个承包商的政策和方法为业主所熟悉并且过去曾证明其能够履行相应的义务。较高的成本可以通过更好的质量，早日建成，并顺利地运营所抵消。

[8] **Limited Competition Construction Bids**

In this method the Owner calls for bids from a few contractors who are known to have specialized in the special type of work and from whom he has, in the past, experienced excellent results. This procedure is usually adopted for private works, where the owner has the right to negotiate directly and to enter into agreement with whomsoever he chooses. This procedure is recommended for the public sector when the work requires specialized knowledge such as industrial construction.

邀请招标

在这种方法中，业主要求少数承包商的投标，这些承包商已被确认专门从事过这种特殊类型的工作，并且在过去已取得了优异的成绩。这种程序通常用于私人工程，其中业主有权直接谈判并与他选择的任何人达成协议。此程序也建议被用于需要工业建筑等专业知识的公共工程。

[9] **Open Competition Construction Bids**

In this method bids are called by public advertisement. The method is usually adopted for public works, as the rules generally require that government and other public contracts be advertised publicly, to obtain the most advantageous terms. Any contractor who is willing to undertake a work and who has the requisite finance and construction equipment to complete it satisfactorily is allowed to submit a bid.

公开招投标

这种方法被称为公开招标。因为规则通常要求公布政府和其他公共合同，这种方法通常用于公共工程以获得最有利的条款。任何愿意承担工程并且拥有足够的资金和施工设备以便于令人满意地完成工程的承包商都可以提交投标书。

[10] The text of a good advertisement should at least include the following information:

—Mode of submitted bids: Bidders should be asked to submit bids in sealed covers, in order to maintain secrecy of quotations.

—Form of bid: It is advisable to get all offers in the same form to facilitate scrutiny, and comparison.

—Name of the inviting authority: This helps the bidder know the persons he will have to deal with and the co-operation he may receive, if his bid is accepted, and such may also affect his quotation.

—Nature of the work and its location: If the nature and location of work is within his operating area, the prospective contractor will want to learn more about the job; else he may not waste time in reading further details.

—Estimated cost of the work: This most briefly indicates the magnitude of the work and enables him to see whether it is too small to interest him or perhaps too large for him to handle.

—Time limit: The time limit within which the work is to be completed should be mentioned. This will be of great help to the bidder to work out a realistic price of the work. The time limit may influence the type and number of construction equipment and workers to be employed, and will vitally affect the contractor's bid.

—The availability of data and forms: where, from whom, at what cost and up to which date blank bid forms and specifications may be obtained and whether a refund will be made upon their return in satisfactory conditions.

—Earnest money required with the bid: The amount and form of security; whether the amount will be accepted in the form of cash, bank guarantee, check or others. Also the number of days within which the amount will be refunded to the unsuccessful bidder.

—Performance security: The amount, form, and procedure of recovering.

—Information regarding drawings: where and when drawings can be examined by bidders. Last date, place and time of receipt of sealed bid.

—The date, time and place and procedure of opening bids.

好的公开招标的文本至少应包括以下信息：

——提交投标方式：应要求投标人以密封的方式提交投标书，以保持报价的保密性。

——投标表单：建议以相同的表单获得所有报价，以便于审查和比较。

——邀请机构的名称：这有助于投标人知道他将和谁打交道和他所能得到的合作，如果投标人的出价被接受，这也可能影响他的报价。

——工程性质及其位置：如果工程的性质和位置在其经营范围内，潜在的承包商将希望了解更多有关工程的信息；否则承包商可能不会浪费时间阅读更多细节；

——估计的工程成本：这最简单地表明了工程的量，使承包商能够清楚它是否太小而不能引起他的兴趣，或者是否太大而无法处理。

——时限：应提及完成工程的时限。这对投标人制订实际工程价格有很大帮助。时限可能会影响施工设备和工人的类型和数量，并将对承包商的投标产生重大影响。

——数据和表格的可用性：可以从何处，从谁，以何种成本和最迟日期以获得空白投标表格和规格，以及是否在满意的条件下退款。

——投标所需的保证金：保证金的数量和形式；是否以现金、银行担保、支票或其他形式接受保证金。此保证金将在多少天内退还给未中标者。

——履约安全：追回的数量、形式和程序。

——有关图纸的信息：投标人可以在何时何地检查图纸。收到密封投标的最后日期、地点和时间。

——开标日期、时间、地点和程序。

References 课文来源

[1] https://en.wikipedia.org
[2] https://www.thenbs.com/knowledge/a-brief-introduction-to-fidic-contracts
[3] http://www.misronet.com/tenders.htm

Unit 3 Building Structure

- ▶ 学习任务　建筑结构
- ▶ 教学时间　2 学时
- ▶ 学习目标　了解建筑结构的主要形式：多层钢结构、砌体结构、框架结构、剪力墙结构以及筒体结构等，初步明了相关的专业术语及其适用范围。

Text 课文

A building or edifice is a structure with a roof and walls standing more or less permanently in one place, such as a house or factory. Buildings come in a variety of sizes, shapes and functions, and have been adapted throughout history for a wide number of factors, from building materials available, to weather conditions, to land prices, ground conditions, specific uses and aesthetic reasons.[1]

Buildings serve several needs of society—primarily as shelter from weather, security, living space, privacy, to store belongings, and to comfortably live and work. A building as a shelter represents a physical division of the human habitat (a place of comfort and safety) and the outside (a place that at times may be harsh and harmful).

Ever since the first cave paintings, buildings have also become objects or canvasses of much artistic expression. In recent years, interest in sustainable planning and building practices has also become an intentional part of the design process of many new buildings.

Single-family residential buildings are most often called houses or homes. Residential buildings containing more than one dwelling unit are called a duplex, apartment building to differentiate them from 'individual' houses. A condominium is an apartment that the occupant owns rather than rents. Houses may also be built in pairs (semi-detached), in terraces where all but two of the houses have others either side; apartments may be built round courtyards or as rectangular blocks surrounded by a piece of ground of varying sizes. Houses which were built as a single dwelling may later be divided into apartments or bedsitters; they may also be converted to another use e. g. an office or a shop.

Building types may range from huts to multi-million dollar high-rise apartment blocks able to house thousands of people. Increasing settlement density in buildings (and smaller distances between buildings) is usually a response to high ground prices resulting from many people wanting to live close to work or similar attractors. Other common building materials are brick, concrete or combinations of either of these with stone.

Residential buildings have different names for their use depending if they are seasonal include holiday cottage (vacation home) or timeshare; size such as a cottage or great house; value such as a shack or mansion; manner of construction such as a log home or mobile home; proximity to the ground such as earth sheltered house, stilt house, or tree house. Also if the residents are in need of special care such as a nursing home, orphanage or prison; or in group housing like barracks or dormitories.

Building structures are classified many forms according to the different materials, such as concrete structure, steel structure and masonry structure. Because the ways that buildings are made to hold up weight, building structures can be divided into different types, such as framed structure, shear wall structure, wall-framed structure, tube structure and so on. [2]

Concrete structure is classified three forms, reinforced concrete, pressed concrete and plain concrete. Reinforced concrete systems are composed of a variety of concrete structural elements that, when synthesized, produce a total system. The components can be broadly classified into: floor slabs, beams, columns, walls and foundations.

3.1 Multi-Story Steel Structural System

Multi-story structural systems mainly apply to office buildings, entertainment centers, large shopping centers and exhibition halls, etc. Multi-story structural systems include columns, beams, and composite decks, roof systems, as well as various alternative wall systems, such as metal, non-metal and curtain walls, etc., and can thus meet people's needs for diverse building styles. [3]

Relative to traditional reinforced concrete structure, the processing of multi-story steel structure system is more convenient to operate with machinery and high precision, and can effectively shorten the construction term and control construction quality.

Steel structure can greatly reduce its dead weight and save ground cost since it has the features of light dead weight and high strength. In addition, heat energy and air conditioning supply are saved because of the reduction of floor slab's thickness and pillar's dimension. [4]

Multi-story steel structure includes girder, secondary beam and column. Hot roll steel or modular welding steel are generally used.

The floor is commonly made of steel floor carrier plate by pouring into reinforced concrete. The steel floor carrier plate has the following features:

—Replacing timber form as the permanent form panel makes the appearance more beautiful;

—Replacing concrete steel structure to resist bending moment. [5]

Multi-story steel structure, as shown in Figure 3-1, has become the first choice of current industrial factory building with the features of variable structures and short construction cycle.

Figure 3-1　Multi-Story Steel Structure

The variable cross section pillar connection system has clear span dimension and there is no need for higher clear mesh, small roof slope, so it can save expenses for heating and heat preservation. The system is an economic design since it can basically meet the requirements of building with various purposes. [6]

The system's inner columns make its span larger, so it doesn't block people and physical distribution. The economic design is applied to the buildings with industrial and commercial purpose, such as office and storage.

The system's roof slope is smaller and its appearance is more beautiful. Simultaneously, it eliminates the useless penthouses; as a result, heating expenses are saved.

The system span can reach as wide as scores of meters because of the existence of inner column. The economic design makes it applied to storehouse and exhibition hall. [7]

The system has the most beautiful roof appearance and smallest roof slope, while the application of truss girder and vertical joist makes the distance between inner columns as long as 12 meters, which is the most ideal choice for shopping center, retail storage and commercial building (shown in Figure 3-2).

The system's roof slope is small with the form of single slope. The increase of inner column increases the inner space and forms a single line of drainage area.[8]

Figure 3-2 E. J. Ourso College of Business

Structural steel is a category of steel used as a construction material for making structural steel shapes. A structural steel shape is a profile, formed with a specific cross section and following certain standards for chemical composition and mechanical properties. Structural steel shapes, sizes, composition, strengths, storage practices, etc., are regulated by standards in most industrialized countries. Steel offers much better compression and tension capacity than concrete and enables lighter construction. Steel structures, as shown in Figure 3-3, use three-dimensional trusses, so they can be larger than reinforced concrete.

Figure 3-3 Steel Structure

3.2 Masonry Structure

Masonry is the building of structures, as shown in Figure 3-4, from individual units, which are often laid in and bound together by mortar; the term masonry can also refer to the units themselves. The common materials of masonry construction are brick, building stone such as marble, granite, travertine, and limestone, cast stone, concrete block and glass block. Masonry is generally a highly durable form of construction.[9] However, the materials used, the quality of the mortar and workmanship, and the pattern in which the units are assembled can substantially affect the durability of the overall masonry construction. A person who constructs masonry is called a mason or bricklayer.

The earliest use of masonry can be traced back to two thousand years ago in China. Before the use of reinforced concrete, masonry such as stone, brick were the main construction materials. Even in the modern time, due to its good quality of heat preservation and easy manufacture and construction, most countries especially developing countries are using it as main material for civil buildings.

Figure 3-4　Masonry Structure

3.3　Framed Structure

Framing, in construction, is the fitting together of pieces to give a structure support and shape. Framing materials are usually wood, engineered wood, or structural steel. The alternative to framed construction is generally called mass wall construction which is made from horizontal layers of stacked materials such as log building, masonry, rammed earth, etc.

Building framing is divided into two broad categories, heavy-frame construction (heavy framing) if the vertical supports are few and heavy such as in timber framing, pole building framing, or steel framing; or light-frame construction (light framing) including balloon, platform and light-steel framing. Light-frame construction using standardized dimensional lumber has become the dominant construction method in North America and Australia because of its economy. Use of minimal structural material allows builders to enclose a large area with minimal cost, while achieving a wide variety of architectural styles.

Structural systems, as shown in Figure 3-5, composed of elements that are long compared with their cross-sectional dimensions are referred to as framed structure, such as beam and column. The elements of a framed structure are defined as linear element since they can transfer the support of loads in only one direction, that is, along the length of the element.[10]

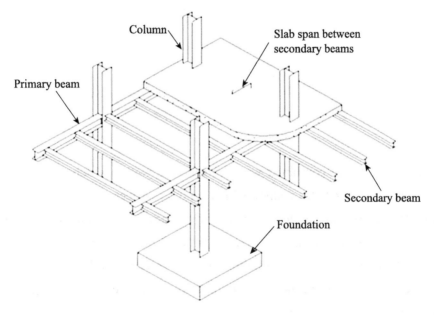

Figure 3-5　Framed Structure

3.4　Shear Wall Structure

In structural engineering, a shear wall is a structural system composed of braced panels (also known as shear panels) to counter the effects of lateral load acting on a structure. Wind and seismic loads are the most common loads that shear walls are designed to carry. A structure of shear walls in the center of a large building—often encasing an elevator shaft or stairwell—form a shear core.

Concrete continuous vertical walls may serve both architecturally as partitions and structurally to carry gravity and lateral loading. In a shear wall structure, such walls are entirely responsible for the lateral load resistance of the building. [11] They act as vertical cantilevers in the form of separate planar walls, and as nonplanar assemblies of connected walls around elevator, stair, and service shafts. Because they are much stiffer horizontally than rigid frames, shear wall structure can be economical to 35 stories (shown in Figure 3-6).

Figure 3-6　Shear Wall Structure with 30 Stories

When shear walls are combined with frames, the walls, which tend to deflect in a flexural configuration, and the frame, which tend to deflect in a shear mode, are constrained to adopt a common deflected shape by the horizontal rigidity of the girder and slabs. As a consequence, the walls and frames interact horizontally, especially at the top, to produce a stiffer and stronger structure. The interacting wall-frame combination, as shown in Figure 3-7, is appropriate for building in the 40 – 60 story range, well beyond that of rigid frames or shear walls alone.

Figure 3-7　Wall-frame Combination

3.5　Tube Structure

In structural engineering, the tube is a system where, to resist lateral loads (wind, seismic, impact), a building is designed to act like a hollow cylinder, cantilevered perpendicular to the ground.[12]

The system can be built using steel, concrete, or composite construction (the discrete use of both steel and concrete). It can be used for office, apartment, and mixed-use buildings. Most buildings of over 40 stories built since the 1960s are of this structural type.

The maximum efficiency of the total structure of a tall building, for both strength and stiffness to resist wind load can be achieved only if all column element can be connected to each other. In such way the entire building acts as a hollow tube or rigid box in the projecting out of the ground, that is tube structure (shown in Figure 3-8).

Figure 3-8　Jinling Hotel Built in 1983 with 108 Metres Height

Glossary 词汇表

edifice ['edɪfɪs] n. 大厦、大建筑物
residential building 住宅建筑
condominium [ˌkɒndə'mɪniəm] n. 公寓
floor slab 楼板
beam [biːm] n. 梁

column ['kɒləm] n. 专栏，圆柱
foundation [faʊn'deɪʃn] n. 基础，地基
concrete structure 混凝土结构
steel structure 钢结构
masonry structure 砌体结构

framed structure 框架结构
shear wall structure 剪力墙结构
wall-framed structure 框架－剪力墙结构
tube structure 筒体结构
multi-story n. 多层
mortar ['mɔːtə(r)] n. 砂浆
marble ['mɑːbl] n. 大理石
granite ['grænɪt] n. 花岗岩
travertine ['trævətɪn] n. 石灰华
limestone ['laɪmstəʊn] n. 石灰岩，石灰石
rammed earth 素土夯实
seismic ['saɪzmɪk] adj. 地震的

Notes 注释

[1] A building or edifice is a structure with a roof and walls standing more or less permanently in one place, such as a house or factory. Buildings come in a variety of sizes, shapes and functions, and have been adapted throughout history for a wide number of factors, from building materials available, to weather conditions, to land prices, ground conditions, specific uses and aesthetic reasons.

建筑物或大厦是一种具有屋顶和墙壁的结构，它或多或少地永久地固定在一个地方，如房屋或工厂。建筑物具有各种尺寸、形状和功能，并在整个历史过程中适应了多种因素，从可用的建筑材料到天气条件、土地价格、地面条件、具体用途和审美原因。

[2] Building structures are classified many forms according to the different materials, such as concrete structure, steel structure and masonry structure. Because the ways that buildings are made to hold up weight, building structures can be divided into different types, such as framed structure, shear wall structure, wall-framed structure, tube structure and so on.

建筑结构根据材料的不同，可分为多种形式，如混凝土结构、钢结构和砌体结构。由于建筑的承重方式不同，建筑结构可分为框架结构、剪力墙结构、框架－剪力墙结构、筒体结构等不同类型。

[3] Multi-story structural systems mainly apply to office buildings, entertainment centers, large shopping centers and exhibition halls, etc. Multi-story structural systems include columns, beams, and composite decks, roof systems, as well as various alternative wall systems, such as metal, non-metal and curtain walls, etc., and can thus meet people's needs for diverse building styles.

多层结构体系多用于写字楼、娱乐中心、大型购物中心及展览厅等。多层结构体系由梁、柱、复合楼面板、屋面体系以及各种可供选择的墙体系统，如金属墙体、非金属墙体、幕墙等组成，可以满足人们对不同建筑体系的需求。

[4] Relative to traditional reinforced concrete structure, the processing of multi-story steel structure system is more convenient to operate with machinery and high precision, and can effectively shorten the construction term and control construction quality.

Steel structure can greatly reduce its dead weight and save ground cost since it has the features of light dead weight and high strength. In addition, heat energy and air conditioning supply are saved because of the reduction of floor slab's thickness and pillar's dimension.

与传统的钢筋混凝土结构相比，多层钢结构体系的加工更容易实现施工机械化、高精度控制，并且可以有效地缩短工期同时确保施工质量。

钢结构能够有效减轻结构自重并节省基础造价，因为它具有自重轻、强度高的特点。另外，板厚和柱子尺寸的减小也降低了建筑耗能。

[5] Multi-story steel structure includes girder, secondary beam and column. Hot roll steel or modular welding steel are generally used.

The floor is commonly made of steel floor carrier plate by pouring into reinforced concrete. The steel floor carrier plate has the following features:

——Replacing timber form as the permanent form panel makes the appearance more beautiful;

——Replacing concrete steel structure to resist bending moment.

多层钢结构由主梁、次梁和柱子组成，它们常由热轧卷钢和焊接型钢构成。楼面板通常由钢板和钢筋混凝土浇筑而成。钢楼面承载板有如下特点：

——替代木模板作为永久性面板使得外观更加美观；

——替代钢筋混凝土抵抗弯矩。

[6] Multi-story steel structure has become the first choice of current industrial factory building with the features of variable structures and short construction cycle.

The variable cross section pillar connection system has clear span dimension and there is no need for higher clear mesh, small roof slope, so it can save expenses for heating and heat preservation. The system is an economic design since it can basically meet the requirements of building with various purposes.

由于结构布置灵活、施工周期短，多层钢结构目前已成为工业厂房的首选。

变截面柱构成的体系尺寸一般为单跨，并且不需要更高的单跨网架和小的屋面坡度，因此它可以节省供热和保温费用。该结构设计是比较经济的，因为它可以基本上满足不同用途的建筑需求。

[7] The system's inner columns make its span larger, so it doesn't block people and physical distribution. The economic design is applied to the buildings with industrial and commercial purpose, such as office and storage.

The system's roof slope is smaller and its appearance is more beautiful. Simultaneously, it eliminates the useless penthouses; as a result, heating expenses are saved.

The system span can reach as wide as scores of meters because of the existence of inner column. The economic design makes it applied to storehouse and exhibition hall.

该体系内部的柱子使得它的跨度可以做到更大，因此它不会阻碍人们和物资调运。经济的设计运用于工业和商业建筑中，例如写字楼和仓库。

该体系的屋面坡度更小使得其外形更加美观。同时，它淘汰了无用的楼顶房间，供热费用得以节省。

内柱的使用使得该体系跨度可达数十米。经济合理的设计使得它适用于仓库和展览厅。

[8] The system has the most beautiful roof appearance and smallest roof slope, while the application of truss girder and vertical joist makes the distance between inner columns as long as 12 meters, which is the most ideal choice for shopping center, retail storage and commercial building.

The system's roof slope is small with the form of single slope. The increase of inner column increases the inner space and forms a single line of drainage area.

该体系拥有最美观的屋顶外形和最小的屋面坡度，同时，桁架梁和立柱的使用使得柱间跨度可达12米，这样的结构是购物中心、零售储存库、商业大厦的最佳选择。

单坡的屋面坡度比较小。利用更多内柱增加了结构内部空间，并且形成单向的排水区。

[9] Masonry is the building of structures from individual units, which are often laid in and bound together by mortar; the term masonry can also refer to the units themselves. The common materials of masonry construction are brick, building stone such as marble, granite, travertine, and limestone, cast stone, concrete block and glass block. Masonry is generally a highly durable form of construction.

砌体结构是由单个砌体所构成的结构，这些砌体通常被铺放并用砂浆砌筑成一个整体。砌筑术语也可以参照其名称本身。砌体工程常见的材料有砖，建筑石材如大理石、花岗岩、石灰华、石灰石、铸石、混凝土块和玻璃块。砌体通常是一种非常耐用的建筑形式。

[10] Structural systems, as shown in Figure 3-5, composed of elements that are long compared with their cross-sectional dimensions are referred to as framed structure, such as beam and column. The elements of a framed, structure are defined as linear element since they can transfer the support of loads in only one direction, that is, along the length of the element.

由与横截面尺寸相比较长的构件组成的结构体系称为框架结构（图3-5），例如梁和柱。框架结构的构件被定义为线性构件，因为它们只能沿着某方向，也就是沿着构件的长度方向传递荷载。

[11] Concrete continuous vertical walls may serve both architecturally as partitions and structurally to carry gravity and lateral loading. In a shear wall structure, such walls are entirely responsible for the lateral load resistance of the building.

混凝土连续垂直墙可以在建筑上用作隔墙并且在结构上用于承载重力和横向荷载。在剪力墙结构中，这些墙完全负责承载建筑物的横向荷载。

[12] In structural engineering, the tube is a system where, to resist lateral loads (wind, seismic, impact), a building is designed to act like a hollow cylinder, cantilevered perpendicular to the ground.

在结构工程中，筒体结构是一种抵抗横向载荷（风、地震、冲击）的系统，建筑物被设计成像一个垂直于地面的悬臂空心圆柱体。

References 课文来源

[1] https://en.wikipedia.org/wiki/Building

Unit 4 Reinforced Concrete

- 学习任务　钢筋混凝土
- 教学时间　3 学时
- 学习目标　初步了解钢筋混凝土发展的历史，掌握混凝土和钢筋的有关技术特点，有利于加深对钢筋混凝土工程施工养护等措施的理解。

Text 课文

Reinforced concrete (RC) is a composite material in which concrete's relatively low tensile strength and ductility are counteracted by the inclusion of reinforcement having higher tensile strength or ductility. The reinforcement is usually, though not necessarily, steel reinforcing bars (rebar) and is usually embedded passively in the concrete before the concrete sets. Reinforcing schemes are generally designed to resist tensile stresses in particular regions of the concrete that might cause unacceptable cracking and/or structural failure. Modern reinforced concrete can contain varied reinforcing materials made of steel, polymers or alternate composite material in conjunction with rebar or not. [1]

For a strong, ductile and durable construction the reinforcement needs to have the following properties at least:

—High relative strength;

—High toleration of tensile strain;

—Good bond to the concrete, irrespective of pH, moisture, and similar factors;

—Thermal compatibility, not causing unacceptable stresses in response to changing temperatures;

—Durability in the concrete environment, irrespective of corrosion or sustained stress for example.

4.1 History

François Coignet, a French industrialist of the nineteenth century, was the first to use iron-reinforced concrete as a technique for constructing building structures. In 1853 he built the first iron reinforced concrete structure, a four-storey house in the suburbs of Paris. In 1854, English builder William B. Wilkinson reinforced the concrete roof and floors in the two-storey house he was constructing. His positioning of the reinforcement demonstrated that, unlike his predecessors, he had knowledge of tensile stresses. G. A. Wayss was a German civil engineer and a pioneer of the iron and steel concrete construction. Ernest L. Ransome was an English-born engineer and early innovator of the reinforced concrete techniques in the end of the 19th century. Ransome's key innovation was to twist the reinforcing steel bar improving bonding with the concrete. Gaining increasing fame from his concrete constructed buildings Ransome was able to build two of the first reinforced concrete bridges in North America. One of the first concrete buildings constructed in the United States, was a private home, designed by William Ward in 1871. The home was designed to be fireproof for his wife. One of the first skyscrapers made with reinforced concrete was the 16-storey Ingalls Building in Cincinnati, constructed in 1904.

4.2 Concrete

Concrete is a composite material composed of coarse aggregate bonded together with a fluid cement that hardens over time. Most concretes used are lime-based concretes such as Portland cement concrete or concretes made with other hydraulic cements, such as ciment fondu. However, asphalt concrete, which is frequently used for road surfaces, is also a type of concrete, where the cement material is bitumen, and polymer concretes are sometimes used where the cementing material is a polymer.

When aggregate is mixed together with dry Portland cement and water, the mixture forms a fluid slurry that is easily poured and molded into shape. The cement reacts chemically with the water and other ingredients to form a hard matrix that binds the materials together into a durable stone-like material that has many uses.[2] Often, additives (such as pozzolans or superplasticizers) are included in the mixture to improve the physical properties of the wet mix or the finished material. Most concrete is poured with reinforcing materials (such as rebar) embedded to provide tensile strength, yielding reinforced concrete.

Famous concrete structures include the Hoover Dam, the Panama Canal, and the Roman Pantheon. The earliest large-scale users of concrete technology were the ancient Romans, and concrete was widely used in the Roman Empire. The Colosseum in Rome was built largely of concrete, and the concrete dome of the Pantheon is the world's largest unreinforced concrete dome. Today, large concrete structures (for example, dams and multi-storey car parks) are usually made with reinforced concrete.

Many types of concrete are available, distinguished by the proportions of the main ingredients below (shown in Table 4-1). In this way or by substitution for the cementitious and aggregate phases, the finished product can be tailored to its application. Strength, density, as well chemical and thermal resistance are variables.

Aggregate consists of large chunks of material in a concrete mix, generally a coarse gravel or crushed rocks such as limestone, or granite, along with finer materials such as sand.

Cement, most commonly Portland cement, is associated with the general term "concrete". A range of other materials can be used as the cement in concrete too. One of the most familiar of these alternative cements is asphalt concrete. Other cementitious materials such as fly ash and slag cement, are sometimes added as mineral admixtures (see Table 4-1)—either preblended with the cement or directly as a concrete component—and become a part of the binder for the aggregate.

To produce concrete from most cements (excluding asphalt), water is mixed with the dry powder and aggregate, which produces a semi-liquid slurry that can be shaped, typically by pouring it into a form. The concrete solidifies and hardens through a chemical process called hydration. The water reacts with the cement, which bonds the other components together, creating a robust stone-like material.

Chemical admixtures are added to achieve varied properties. These ingredients may accelerate or slow down the rate at which the concrete hardens, and impart many other useful properties including increased tensile strength, entrainment of air, and/or water resistance. Reinforcement is often included in concrete.

Mineral admixtures are becoming more popular in recent decades. The use of recycled materials as concrete ingredients has been gaining popularity because of increasingly stringent environmental legislation, and the discovery that such materials often have complementary and valuable properties. The most conspicuous of these are fly ash, a by-product of coal-fired power plants, ground granulated blast furnace slag, a byproduct of steelmaking, and silica fume, a byproduct of industrial electric arc furnaces. The use of these materials in concrete reduces the amount of resources required, as the mineral admixtures act as a partial cement replacement.

Table 4-1 Components of Cement Comparison of Chemical and Physical Characteristics[a]

Property	Portland Cement	Siliceous (ASTM C618 Class F) Fly Ash	Calcareous (ASTM C618 Class C) Fly ash	Slag Cement	Silica Fume
SiO_2 content (%)	21.9	52	35	35	85~97
Al_2O_3 content (%)	6.9	23	18	12	—
Fe_2O_3 content (%)	3	11	6	1	—
CaO content (%)	63	5	21	40	<1
MgO content (%)	2.5	—	—	—	—
SO_3 content (%)	1.7	—	—	—	—
Specific surface[b] / (m^2/kg)	370	420	420	400	15,000~30,000
Specific gravity	3.15	2.38	2.65	2.94	2.22
General use in concrete	Primary binder	Cement replacement	Cement replacement	Cement replacement	Property enhancer

[a] Values shown are approximate; those of a specific material may vary.

[b] Specific surface measurements for silica fume by nitrogen adsorption (BET) method, others by air permeability method (Blaine).

Concrete production is the process of mixing together the various ingredients—water, aggregate, cement, and any additives—to produce concrete. Concrete production is time-sensitive. Once the ingredients are mixed, workers must put the concrete in place before it hardens. In modern usage, most concrete production takes place in a large type of industrial facility called a concrete plant, or often a batch plant.

In general usage, concrete plants come in two main types, ready mix plants and central mix plants. A ready mix plant mixes all the ingredients except water, while a central mix plant mixes all the ingredients including water. A central mix plant offers more accurate control of the concrete quality through better measurements of the amount of water added, but must be placed closer to the work site where the concrete will be used, since hydration begins at the plant.

A concrete plant consists of large storage hoppers for various reactive ingredients like cement, storage for bulk ingredients like aggregate and water, mechanisms for the addition of various additives and amendments, machinery to accurately weigh, move, and mix some or all of those ingredients, and facilities to dispense the mixed concrete, often to a concrete mixer truck.

Modern concrete is usually prepared as a viscous fluid, so that it may be poured into forms, which are containers erected in the field to give the concrete its desired shape. Concrete formwork can be prepared in several ways, such as Slip forming and Steel plate construction. Alternatively, concrete can be mixed into dryer, non-fluid forms and used in factory settings to manufacture Precast concrete products.

A wide variety of equipment is used for processing concrete, from hand tools to heavy industrial machinery. Whichever equipment builders use, however, the objective is to produce the desired building material; ingredients must be properly mixed, placed, shaped, and retained within time constraints. Any interruption in pouring the concrete can cause the initially placed material to begin to set before the next batch is added on top. This creates a horizontal plane of weakness called a cold joint between the two batches. Once the mix is where it should be, the curing process must be controlled to ensure that the concrete attains the desired attributes.[3] During concrete preparation, various technical details may affect the quality and nature of the product.

When initially mixed, Portland cement and water rapidly form a gel of tangled chains of interlocking crystals, and components of the gel continue to react over time. Initially the gel is fluid, which improves workability and aids in placement of the material, but as the concrete sets, the chains of crystals join into a rigid structure, counteracting the fluidity of the gel and fixing the particles of aggregate in place. During curing, the cement continues to react with the residual water in a process of hydration. In properly formulated concrete, once this curing process has terminated the product has the desired physical and chemical properties. Among the qualities typically desired, are mechanical strength, low moisture permeability, and chemical and volumetric stability.

Curing is the hydration process that occurs after the concrete has been placed. In chemical terms, curing allows calcium-silicate hydrate (C-S-H) to form. To gain strength and harden fully, concrete curing requires time. Hydration and hardening of concrete during the first three days is critical. Abnormally fast drying and shrinkage due to factors such as evaporation from wind during placement may lead to increased tensile stresses at a time when it has not yet gained sufficient strength, resulting in greater shrinkage cracking. In around 4 weeks, typically over 90% of the final strength is reached, although strengthening may continue for decades. Properly curing concrete leads to increased strength and lower permeability and avoids cracking where the surface dries out prematurely. Care must also be taken to avoid freezing or overheating due to the exothermic setting of cement. Improper curing can cause scaling, reduced strength, poor abrasion resistance and cracking.

4.3 Steel Works

Rebar (short for reinforcing bar, as shown in Figure 4-1), collectively known as reinforcing steel and reinforcement steel, is a steel bar or mesh of steel wires (shown in Table 4-2 and Table 4-3) used as a tension device in reinforced concrete and reinforced masonry structures to strengthen and hold the concrete in tension. Rebar's surface is often patterned to form a better bond with the concrete.[4]

a)

b)

Figure 4-1 Steel Works in Foundation

Table 4-2 US Rebar Size Chart

Imperial bar size	Metric size	Linear Mass Density		Nominal diameter		Nominal area	
		lb/ft	/(kg/m)	/in	/mm	/in²	/mm²
#2	No. 6	0.167	0.249	0.250=1/4	6.35	0.05	32
#3	No. 10	0.376	0.561	0.375=3/8	9.525	0.11	71
#4	No. 13	0.668	0.996	0.500=1/2	12.7	0.20	129
#5	No. 16	1.043	1.556	0.625=5/8	15.875	0.31	200
#6	No. 19	1.502	2.24	0.750=3/4	19.05	0.44	284
#7	No. 22	2.044	3.049	0.875=7/8	22.225	0.60	387

(续)

Imperial bar size	Metric size	Linear Mass Density		Nominal diameter		Nominal area	
		lb/ft	/(kg/m)	/in	/mm	/in^2	/mm^2
#8	No.25	2.670	3.982	1.000	25.4	0.79	509
#9	No.29	3.400	5.071	1.128	28.65	1.00	645
#10	No.32	4.303	6.418	1.270	32.26	1.27	819
#11	No.36	5.313	7.924	1.410	35.81	1.56	1006
#14	No.43	7.650	11.41	1.693	43	2.25	1452
#18	No.57	13.60	20.284	2.257	57.3	4.00	2581
#18J		14.60	21.775	2.337	59.4	4.29	2678

Table 4-3 US Rebar Size Chart

Metric bar size	Linear Mass Density/(kg/m)	Nominal diameter/mm	Cross-sectional Area/mm^2
6.0	0.222	6	28.3
8.0	0.395	8	50.3
10.0	0.617	10	78.5
12.0	0.888	12	113
14.0	1.21	14	154
16.0	1.58	16	201
20.0	2.47	20	314
25.0	3.85	25	491
28.0	4.83	28	616
32.0	6.31	32	804
40.0	9.86	40	1257
50.0	15.4	50	1963

Steel has a thermal expansion coefficient nearly equal to that of modern concrete. If this were not so, it would cause problems through additional longitudinal and perpendicular stresses at temperatures different from the temperature of the setting. Although rebar has ribs that bind it mechanically to the concrete, it can still be pulled out of the concrete under high stresses, an occurrence that often accompanies a larger-scale collapse of the structure. To prevent such a failure, rebar is either deeply embedded into adjacent structural members (40-60 times the diameter), or bent and hooked at the ends to lock it around the concrete and other rebar. This first approach increases the friction locking the bar into place, while the second makes use of the high compressive strength of concrete.

Common rebar is made of unfinished tempered steel, making it susceptible to rusting.

Normally the concrete cover is able to provide a pH value higher than 12 avoiding the corrosion reaction. Too little concrete cover can compromise this guard through carbonation from the surface, and salt penetration. Too much concrete cover can cause bigger crack widths which also compromises the local guard. As rust takes up greater volume than the steel from which it was formed, it causes severe internal pressure on the surrounding concrete, leading to cracking, spalling, and ultimately, structural failure. This phenomenon is known as oxide jacking. This is a particular problem where the concrete is exposed to salt water, as in bridges where salt is applied to roadways in winter, or in marine applications. Uncoated, corrosion-resistant low carbon/chromium (microcomposite), epoxy-coated, galvanized or stainless steel rebars may be employed in these situations at greater initial expense, but significantly lower expense over the service life of the project. Extra care is taken during the transport, fabrication, handling, installation, and concrete placement process when working with epoxy-coated rebar, because damage will reduce the long-term corrosion resistance of these bars. Even damaged bars have shown better performance than uncoated reinforcing bars, though issues from debonding of the epoxy coating from the bars and corrosion under the epoxy film have been reported.

4.4 Behavior of Reinforced Concrete

Concrete is a mixture of coarse (stone or brick chips) and fine (generally sand or crushed stone) aggregates with a paste of binder material (usually Portland cement) and water. When cement is mixed with a small amount of water, it hydrates to form microscopic opaque crystal lattices encapsulating and locking the aggregate into a rigid structure. The aggregates used for making concrete should be free from harmful substances like organic impurities, silt, clay, lignite etc. Typical concrete mixes have high resistance to compressive stresses [about 4000 psi (28 MPa)]; however, any appreciable tension (e.g., due to bending) will break the microscopic rigid lattice, resulting in cracking and separation of the concrete. For this reason, typical non-reinforced concrete must be well supported to prevent the development of tension.

If a material with high strength in tension, such as steel, is placed in concrete, then the composite material, reinforced concrete, resists not only compression but also bending and other direct tensile actions. A reinforced concrete section where the concrete resists the compression and steel resists the tension can be made into almost any shape and size for the construction industry.

Three physical characteristics give reinforced concrete its special properties:

1) The coefficient of thermal expansion of concrete is similar to that of steel, eliminating large internal stresses due to differences in thermal expansion or contraction.

2) When the cement paste within the concrete hardens, this conforms to the surface details of the steel, permitting any stress to be transmitted efficiently between the different materials. Usually steel bars are roughened or corrugated to further improve the bond or cohesion between the concrete and steel.

3) The alkaline chemical environment provided by the alkali reserve (KOH, NaOH) and the portlandite (calcium hydroxide) contained in the hardened cement paste causes a passivating film to form on the surface of the steel, making it much more resistant to corrosion than it would be in neutral or acidic conditions. When the cement paste is exposed to the air and meteoric water reacts with the atmospheric CO_2, portlandite and the calcium silicate hydrate (CSH) of the hardened cement paste become progressively carbonated and the high PH gradually decreases from 13.5 – 12.5 to 8.5, the PH of water in equilibrium with calcite (calcium carbonate) and the steel is no longer passivated.

4) The relative cross-sectional area of steel required for typical reinforced concrete is usually quite small and varies from 1% for most beams and slabs to 6% for some columns. Reinforcing bars are normally round in cross-section and vary in diameter. Reinforced concrete structures sometimes have provisions such as ventilated hollow cores to control their moisture & humidity.

5) The reinforcement in a RC structure, such as a steel bar, has to undergo the same strain or deformation as the surrounding concrete in order to prevent discontinuity, slip or separation of the two materials under load. Maintaining composite action requires transfer of load between the concrete and steel.[5] The direct stress is transferred from the concrete to the bar interface so as to change the tensile stress in the reinforcing bar along its length; this load transfer is achieved by means of bond (anchorage) and is idealized as a continuous stress field that develops in the vicinity of the steel-concrete interface.

Glossary 词汇表

reinforced concrete 钢筋混凝土
tensile strength 抗拉强度
ductility [dʌk'tɪləti] n. 韧性，塑性
corrosion [kə'rəʊʒn] n. 腐蚀
innovation [ˌɪnə'veɪʃn] n. 改革，创新

aggregate ['ægrɪgət] n. 骨料，集料
cement [sɪ'ment] n. 水泥
slurry ['slʌri] n. 泥浆
ingredient [ɪn'griːdiənt] n. 成分
additive ['ædətɪv] n. 添加剂

admixture [əd'mɪkstʃə(r)] n. 混合物
legislation [ˌledʒɪs'leɪʃn] n. 法律，法规
conspicuous [kən'spɪkjuəs] adj. 明显的
hopper ['hɒpə(r)] n. 储料器
concrete formwork 混凝土模板作业
calcium-silicate hydrate 硅酸钙水合物
hydration [haɪ'dreɪʃn] n. 水化作用
curing ['kjʊərɪŋ] n. 养护
scaling ['skeɪlɪŋ] n. 起皮，鳞片状剥落
abrasion [ə'breɪʒn] n. 磨蚀
cracking ['krækɪŋ] n. 开裂
spalling ['spɔːlɪŋ] n. 剥落
rebar [rɪ'bɑː] n. 加强筋，钢筋，螺纹钢筋

marine [mə'riːn] adj. 海生的；海产的；航海的，海运的
epoxy-coated n. 环氧树脂涂层
galvanized ['gælvənaɪzd] adj. 镀锌的
transport ['trænspɔːt] vt. 运送，运输
fabrication [ˌfæbrɪ'keɪʃn] n. 制造
handling ['hændlɪŋ] n. 处理
installation [ˌɪnstə'leɪʃn] n. 安装，设置
moisture ['mɔɪstʃə(r)] n. 水分，湿气，潮湿，降雨量
humidity [hjuː'mɪdəti] n. （空气中的）湿度，潮湿，[物] 湿度
steel bar 钢筋

Notes 注释

[1] Reinforced concrete (RC) is a composite material in which concrete's relatively low tensile strength and ductility are counteracted by the inclusion of reinforcement having higher tensile strength or ductility. The reinforcement is usually, though not necessarily, steel reinforcing bars (rebar) and is usually embedded passively in the concrete before the concrete sets. Reinforcing schemes are generally designed to resist tensile stresses in particular regions of the concrete that might cause unacceptable cracking and/or structural failure. Modern reinforced concrete can contain varied reinforcing materials made of steel, polymers or alternate composite material in conjunction with rebar or not.

钢筋混凝土（RC）是一种通过具有较高拉伸强度或延展性的钢筋来抵消混凝土相对较低的拉伸强度和延展性的复合材料。虽然有例外，加固材料一般为钢筋，钢筋通常在混凝土凝固前被动地植入混凝土中。加固方案一般是为了抵抗混凝土中可能引起不可接受的开裂和/或结构破坏的特定区域的拉应力。现代钢筋混凝土可以包含不同的加强材料如钢筋、聚合物或替代的复合材料与钢筋联合制成的增强材料。

[2] When aggregate is mixed together with dry Portland cement and water, the mixture forms a fluid slurry that is easily poured and molded into shape. The cement reacts chemically with the water and other ingredients to form a hard matrix that binds the materials together into a durable stone-like material that has many uses.

当骨料与干燥的波特兰水泥和水混合在一起时，混合物形成容易浇筑成型的流体浆料。水泥与水和其他成分进行化学反应，形成将材料结合在一起的坚硬基体，成为具有许多用途的耐用的石质材料。

[3] A wide variety of equipment is used for processing concrete, from hand tools to heavy industrial machinery. Whichever equipment builders use, however, the objective is to produce the desired building material; ingredients must be properly mixed, placed, shaped, and retained within time constraints. Any interruption in pouring the concrete can cause the initially placed material to begin to set before the next batch is added on top. This creates a horizontal plane of weakness called a cold joint between the two batches. Once the mix is where it should be, the curing process must be controlled to ensure that the concrete attains the desired attributes.

从手工工具到重型工业机械，用于加工混凝土的设备种类繁多。然而，无论建造者使用哪种设备，其目标是生产所需的建筑材料，配料必须在时间限制内适当混合、放置、成型和保留。浇筑混凝土时的任何中断都可能导致最初放置的材料在下一批加在顶部之前开始凝固。这在两个批次之间形成了一个薄弱的水平面，称为冷接头。一旦混合物达到应有的水平，必须控制固化过程，以确保混凝土达到所需的性能。

[4] Rebar (short for reinforcing bar, as shown in Figure 4-1), collectively known as reinforcing steel and reinforcement steel, is a steel bar or mesh of steel wires used as a tension device in reinforced concrete and reinforced masonry structures to strengthen and hold the concrete in tension. Rebar's surface is often patterned to form a better bond with the concrete.

加强筋（简称钢筋，图 4-1），统称为钢加强筋，是钢筋或钢丝网，在钢筋混凝土和配筋砌体结构中作为一种张力装置，用来加强和保持混凝土处于张力状态。钢筋表面通常轧纹以便与混凝土形成更好的粘结。

[5] The reinforcement in a RC structure, such as a steel bar, has to undergo the same strain or deformation as the surrounding concrete in order to prevent discontinuity, slip or separation of the two materials under load. Maintaining composite action requires transfer of load between the concrete and steel.

钢筋混凝土结构中的钢加强筋，如钢筋，必须承受与周围混凝土相同的应变或变形，以防止这两种材料在荷载作用下产生不连续、滑移或分离。保持复合作用需要在混凝土和钢筋之间传递荷载。

References 课文来源

[1] https://en.wikipedia.org/wiki/Concrete
[2] https://en.wikipedia.org/wiki/Rebar

Unit 5 Prestressed Concrete

> ▸ 学习任务　预应力混凝土
> ▸ 教学时间　2 学时
> ▸ 学习目标　了解预应力混凝土的先张法和后张法的有关工艺以及预应力混凝土的养护技术要点。

Text 课文

Prestressed concrete is a concrete construction material which is placed under compression prior to supporting any applied loads (i. e. it is "pre" stressed). A more technical definition is "Structural concrete in which internal stresses have been introduced to reduce potential tensile stresses in the concrete resulting from loads." This compression is produced by the tensioning of high-strength "tendons" located within or adjacent to the concrete volume, and is done to improve the performance of the concrete in service. Tendons may consist of single wires, multi-wire strands or threaded bars, and are most commonly made from high-tensile steels, carbon fibre or aramid fibre. The essence of prestressed concrete is that once the initial compression has been applied, the resulting material has the characteristics of high-strength concrete when subject to any subsequent compression forces, and of ductile high-strength steel when subject to tension forces.

Prestressed concrete is used in a wide range of building and civil structures where its improved concrete performance can allow longer spans, reduced structural thicknesses, and material savings to be realized compared to reinforced concrete. Typical applications range through high-rise buildings, residential slabs, foundation systems, bridge and dam structures, silos and tanks, industrial pavements and nuclear containment structures.[1]

Tensioning (or "stressing") of the tendons may be undertaken either before (pre-tensioning) or after (post-tensioning) the concrete itself is cast. Tendons may be located either within the concrete volume (internal prestressing), or wholly outside of it (external prestressing).

Whereas pre-tensioned concrete by definition uses tendons directly bonded to the concrete, post-tensioned concrete can use either bonded or unbonded tendons. Tensioning systems can be classed as either monostrand systems, where each tendon's strand or wire is stressed individually, or multi-strand systems where all strands or wires in a tendon are stressed simultaneously.

5.1 Pre-tensioned Concrete

Pre-tensioned concrete, as shown in Figure 5-1, is most commonly used for the fabrication of structural beams, floor slabs, hollow-core planks, balconies, lintels, driven piles, water tanks and concrete pipes. Pre-tensioned concrete is a variant of prestressed concrete where the tendons are tensioned prior to the concrete being cast. The concrete bonds to the tendons as it cures, following which the end-anchoring of the tendons is released, and the tendon tension forces are transferred to the concrete as compression by static friction.[2]

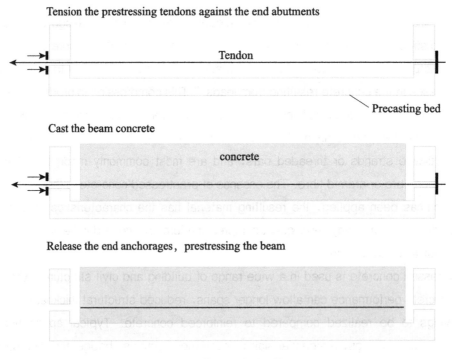

Figure 5-1　Pre-tensioning Process

Pre-tensioning is a common prefabrication technique, where the resulting concrete element is manufactured remotely from the final structure location and transported to site once cured. It requires strong, stable end-anchorage points between which the tendons are stretched. These anchorages form the ends of a "casting bed" which may be many times the length of the concrete element being fabricated. This allows multiple elements to be constructed end-on-end in the one pre-tensioning operation, allowing significant productivity benefits and economies of scale to be realized for this method of construction.

The amount of bond (or adhesion) achievable between the freshly set concrete and the surface of the tendons is critical to the pre-tensioning process, as it determines when the tendon anchorages can be safely released. Higher bond strength in early-age concrete allows more economical fabrication as it speeds production. To promote this, pre-tensioned tendons are usually composed of isolated single wires or strands, as this provides a greater surface area for bond action than bundled strand tendons.

Unlike those of post-tensioned concrete, the tendons of pre-tensioned concrete elements generally form straight lines between end-anchorages. Where "profiled" or "harped" tendons are required, one or more intermediate deviators are located between the ends of the tendon to hold the tendon to the desired non-linear alignment during tensioning. Such deviators usually act against substantial forces, and hence require a robust casting bed foundation system. Straight tendons are typically used in "linear" precast elements such as shallow beams, hollow-core planks and slabs, whereas profiled tendons are more commonly found in deeper precast bridge beams and girders.

5.2 Post-tensioned Concrete

Post-tensioned concrete, as shown in Figure 5-2, is a variant of prestressed concrete where the tendons are tensioned after the surrounding concrete structure has been cast. The tendons are not placed in direct contact with the concrete, but are encapsulated within a protective sleeve or duct which is either cast into the concrete structure or placed adjacent to it. At each end of a tendon is an anchorage assembly firmly fixed to the surrounding concrete. Once the concrete has been cast and set, the tendons are tensioned ("stressed") by pulling the tendon ends through the anchorages while pressing against the concrete.[3] The large forces required to tension the tendons result in a significant permanent compression being applied to the concrete once the tendon is "locked-off" at the anchorage. The method of locking the tendon-ends to the anchorage is dependent upon the tendon composition, with the most common systems being "button-head" anchoring (for wire tendons), split-wedge anchoring (for strand tendons), and threaded anchoring (for bar tendons).

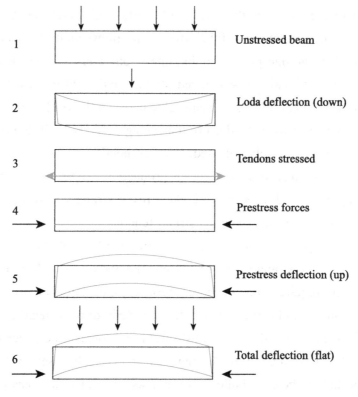

Figure 5-2 Post-tensioning Process

Tendon encapsulation systems are constructed from plastic or galvanized steel materials, and are classified into two main types: those where the tendon element is subsequently bonded to the surrounding concrete by internal grouting of the duct after stressing (bonded post-tensioning); and those where the tendon element is permanently debonded from the surrounding concrete, usually by means of a greased sheath over the tendon strands (unbonded post-tensioning).

Casting the tendon ducts/sleeves into the concrete before any tensioning occurs allows them to be readily "profiled" to any desired shape including incorporating vertical and/or horizontal curvature. When the tendons are tensioned, this profiling results in reaction forces being imparted onto the hardened concrete, and these can be beneficially used to counter any loadings subsequently applied to the structure.

5.2.1 Bonded post-tensioning

Bonded post-tensioning has prestressing tendons permanently bonded to the surrounding concrete by the in situ grouting of their encapsulating ducting following tendon tensioning. This grouting is undertaken for three main purposes: to protect the tendons against corrosion; to permanently "lock-in" the tendon pre-tension, thereby removing the long-term reliance upon the end-anchorage systems; and to improve certain structural behaviours of the final concrete structure.[4]

Bonded post-tensioning characteristically uses tendons each comprising bundles of elements (e. g. strands or wires) placed inside a single tendon duct, with the exception of bars which are mostly used unbundled. This bundling make for more efficient tendon installation and grouting processes, since each complete tendon requires only one set of end-anchorages and one grouting operation. Ducting is fabricated from a durable and corrosion-resistant material such as plastic (e. g. polyethylene) or galvanised steel, and can be either round or rectangular/oval in cross-section. The tendon sizes used are highly dependent upon the application, ranging from building works typically using between 2-strands and 6-strands per tendon, to specialised dam works using up to 91-strands per tendon.

Fabrication of bonded tendons is generally undertaken on-site, commencing with the fitting of end-anchorages to formwork, placing the tendon ducting to the required curvature profiles, and reeving (or threading) the strands or wires through the ducting. Following concreting and tensioning, the ducts are pressure-grouted and the tendon stressing-ends sealed against corrosion.

5.2.2 Unbonded post-tensioning

Unbonded post-tensioning differs from bonded post-tensioning by allowing the tendons permanent freedom of longitudinal movement relative to the concrete. This is most commonly achieved by encasing each individual tendon element within a plastic sheathing filled with a corrosion-inhibiting grease, usually lithium based. Anchorages at each end of the tendon transfer the tensioning force to the concrete, and are required to reliably perform this role for the life of the structure.[5]

Unbonded post-tensioning can take the form of: Individual strand tendons placed directly into the concreted structure (e. g. buildings, ground slabs); or Bundled strands, individually greased-and-sheathed, forming a single tendon within an encapsulating duct that is placed either within or adjacent to the concrete (e. g. restressable anchors, external post-tensioning).

For individual strand tendons, no additional tendon ducting is used and no post-stressing grouting operation is required, unlike for bonded post-tensioning. Permanent corrosion protection of the strands is provided by the combined layers of grease, plastic sheathing, and surrounding concrete. Where strands are bundled to form a single unbonded tendon, an enveloping duct of plastic or galvanised steel is used and its interior free-spaces grouted after stressing. In this way, additional corrosion protection is provided via the grease, plastic sheathing, grout, external sheathing, and surrounding concrete layers.

Individually greased-and-sheathed tendons are mostly fabricated off-site by an extrusion process. The bare steel strand is fed into a greasing chamber and then passed to an extrusion unit where molten plastic forms a continuous outer coating. Finished strands can be cut-to-length and fitted with "dead-end" anchor assemblies as required for the project.

5.3 Tendon Durability and Corrosion Protection

Long-term durability is an essential requirement for prestressed concrete given its significance as a ubiquitous, modern construction material. Research on the durability performance of in-service prestressed structures has been undertaken since the 1960s, and anti-corrosion technologies for tendon protection have been continually improved since the earliest systems were developed.

The durability of prestressed concrete is principally determined by the level of corrosion protection provided to any high-strength steel elements within the prestressing tendons. Also critical is the protection afforded to the end-anchorage assemblies of unbonded tendons or cable-stay systems, as the anchorages of both of these are required to retain the prestressing forces permanently.[6] Failure of any of these components can result in the release of prestressing forces, or the physical rupture of stressing tendons.

Modern prestressing systems deliver long-term durability by addressing the following areas:

(1) Tendon grouting (bonded tendons)

Bonded tendons consist of bundled high-strength (steel) strands placed inside ducts located within the surrounding concrete. To ensure full protection to these strands, the ducts must be pressure-filled with a corrosion-inhibiting grout, without leaving any voids, following strand-tensioning.

(2) Tendon coating (unbonded tendons)

Unbonded tendons comprise individual high-strength strands coated in an anti-corrosion grease or wax, and fitted with a durable plastic-based full-length sleeve or sheath. The sleeving is required to be undamaged over the tendon length, and it must extend fully into the anchorage fittings at each end of the tendon.

(3) Double-layer encapsulation

Prestressing tendons requiring permanent monitoring and/or force adjustment, such as stay-cables and re-stressable dam anchors, will typically employ double-layer corrosion protection. Such tendons are composed of individual strands, grease-coated and sleeved, collected into a strand-bundle and placed inside encapsulating polyethylene outer ducting.

The remaining void space within the duct is pressure-grouted, providing a multi-layer polythene-grout-plastic-grease protection barrier system for each strand.

(4) Anchorage protection

In all post-tensioned installations, protection of the end-anchorages against corrosion is essential, and critically so for unbonded systems.

Glossary 词汇表

prestressed concrete 预应力混凝土
tensioning ['tenʃnɪŋ] n. 张拉
tendon ['tendən] n. 筋,索
pre-tensioning n. 先张法
prefabrication [ˌpriːfæbrɪ'keɪʃn] n. 预制
end-anchorage n. 端部锚固
plank [plæŋk] n. (厚)木板
girder ['ɡɜːdə(r)] n. 大梁
post-tensioning n. 后张法
wire tendon 钢丝束
strand tendon 钢绞线
bar tendon 钢筋
bonded post-tensioning 有粘结后张法
unbonded post-tensioning 无粘结后张法
ducting ['dʌktɪŋ] n. 管道
grouting ['ɡraʊtɪŋ] n. 灌浆
coating ['kəʊtɪŋ] n. 涂层
encapsulation [ɪnˌkæpsjuˈleɪʃən] n. 封装

Notes 注释

[1] Prestressed concrete is used in a wide range of building and civil structures where its improved concrete performance can allow longer spans, reduced structural thicknesses, and material savings to be realized compared to reinforced concrete. Typical applications range through high-rise buildings, residential slabs, foundation systems, bridge and dam structures, silos and tanks, industrial pavements and nuclear containment structures.

预应力混凝土广泛应用于建筑和土木结构中，与钢筋混凝土相比，其改进的混凝土性能可以允许更大的跨度，降低结构的厚度，以及节省材料。典型的应应用范围包括高层建筑、住宅楼板、地基基础、桥梁和水坝结构、筒仓和贮槽、工业路面铺装和核安全壳结构。

[2] Pre-tensioned concrete, as shown in Figure 5-1, is most commonly used for the fabrication of structural beams, floor slabs, hollow-core planks, balconies, lintels, driven piles, water tanks and concrete pipes. Pre-tensioned concrete is a variant of prestressed concrete where the tendons are tensioned prior to the concrete being cast.

The concrete bonds to the tendons as it cures, following which the end-anchoring of the tendons is released, and the tendon tension forces are transferred to the concrete as compression by static friction.

先张法混凝土（图5-1）最常用来制作结构梁、楼板、空心板、阳台、过梁、打入桩、水箱和混凝土管。先张法混凝土是预应力混凝土的一种变体，预应力筋在混凝土浇筑之前就被拉紧。混凝土在养护固化时粘结到预应力筋上，随后松开预应力筋的末端锚固，张拉力通过静摩擦力转移到混凝土上，使其产生压缩作用。

[3] Post-tensioned concrete, as shown in Figure 5-2, is a variant of prestressed concrete where the tendons are tensioned after the surrounding concrete structure has been cast. The tendons are not placed in direct contact with the concrete, but are encapsulated within a protective sleeve or duct which is either cast into the concrete structure or placed adjacent to it. At each end of a tendon is an anchorage assembly firmly fixed to the surrounding concrete. Once the concrete has been cast and set, the tendons are tensioned ("stressed") by pulling the tendon ends through the anchorages while pressing against the concrete.

后张法混凝土（图5-2）是预应力混凝土的一种变体，当周围的混凝土结构被浇筑后，预应力筋就被拉紧了。预应力筋不与混凝土直接接触，而是被封装在一个保护套管或管道中，保护套管或管道被浇筑到混凝土结构中或与其相邻放置。在预应力筋的两端是固定在周围混凝土上的锚具组件。一旦混凝土已浇筑固化，张拉预应力筋后通过端部锚具使混凝土产生压缩作用。

[4] Bonded post-tensioning has prestressing tendons permanently bonded to the surrounding concrete by the in situ grouting of their encapsulating ducting following tendon tensioning. This grouting is undertaken for three main purposes: to protect the tendons against corrosion; to permanently "lock-in" the tendon pre-tension, thereby removing the long-term reliance upon the end-anchorage systems; and to improve certain structural behaviours of the final concrete structure.

有粘结后张法中，通过在预应力筋张紧之后的封装管道进行原位灌浆，预应力筋和周围的混凝土永久粘结。这种注浆有三个主要目的：保护预应力筋免受腐蚀；永久"锁定"预应力筋预紧，从而消除对端部锚固系统的长期依赖；提高最终混凝土结构的某些结构性能。

[5] Unbonded post-tensioning differs from bonded post-tensioning by allowing the tendons permanent freedom of longitudinal movement relative to the concrete. This is most commonly achieved by encasing each individual tendon element within a plastic sheathing filled with a corrosion-inhibiting grease, usually lithium based. Anchorages at each end of the tendon transfer the tensioning force to the concrete, and are

required to reliably perform this role for the life of the structure.

无粘结后张法不同于有粘结后张法，它允许预应力筋相对于混凝土具有纵向运动的永久自由度，这通常通过将每个单独的预应力筋封装在填充有防腐润滑脂（通常为锂基）的塑料外壳中来实现。预应力筋两端的锚固装置将拉力转移到混凝土，并且要求在结构的使用寿命期间可靠地发挥这种作用。

[6] The durability of prestressed concrete is principally determined by the level of corrosion protection provided to any high-strength steel elements within the prestressing tendons. Also critical is the protection afforded to the end-anchorage assemblies of unbonded tendons or cable-stay systems, as the anchorages of both of these are required to retain the prestressing forces permanently.

预应力混凝土的耐久性主要取决于提供给预应力筋内任何高强度钢构件的防腐蚀水平。同样重要的是，对无粘结预应力筋或拉索系统的端部锚固组件提供保护，因为这两者的端部锚固组件也需要永久地保持预应力。

References 课文来源

[1] https://en.wikipedia.org/wiki/Prestressed_concrete

Unit 6 Geotechnical Engineering

- 学习任务　岩土工程
- 教学时间　2 学时
- 学习目标　了解岩土工程的相关术语，重点掌握浅基础和深基础的相关技术特点。

Text 课文

Geotechnical engineering is the branch of civil engineering concerned with the engineering behavior of earth materials. Geotechnical engineering is important in civil engineering, but also has applications in military, mining, petroleum and other engineering disciplines that are concerned with construction occurring on the surface or within the ground.[1] Geotechnical engineering uses principles of soil mechanics and rock mechanics to investigate subsurface conditions and materials; determine the relevant physical/mechanical and chemical properties of these materials; evaluate stability of natural slopes and man-made soil deposits; assess risks posed by site conditions; design earthworks and structure foundations; and monitor site conditions, earthwork and foundation construction.

A typical geotechnical engineering project begins with a review of project needs to define the required material properties. Then follows a site investigation of soil, rock, fault distribution and bedrock properties on and below an area of interest to determine their engineering properties including how they will interact with, on or in a proposed construction. Site investigations are needed to gain an understanding of the area in or on which the engineering will take place. Investigations can include the assessment of the risk to humans, property and the environment from natural hazards such as earthquakes, landslides, sinkholes, soil liquefaction, debris flows and rockfalls.

A geotechnical engineer then determines and designs the type of foundations, earthworks, and/or pavement subgrades required for the intended man-made structures to be built. Foundations are designed and constructed for structures of various sizes such as high-rise buildings, bridges, medium to large commercial buildings, and smaller structures where the soil conditions do not allow code-based design.

Foundations built for above-ground structures include shallow and deep foundations. Retaining structures include earth-filled dams and retaining walls. Earthworks include embankments, tunnels, dikes and levees, channels, reservoirs, deposition of hazardous waste and sanitary landfills.

Geotechnical engineering is also related to coastal and ocean engineering. Coastal engineering can involve the design and construction of wharves, marinas, and jetties. Ocean engineering can involve foundation and anchor systems for offshore structures such as oil platforms.

6.1 History of Geotechnical Engineering

Humans have historically used soil as a material for flood control, irrigation purposes, burial sites, building foundations, and as construction material for buildings. Until the 18th century, however, no theoretical basis for soil design had been developed and the discipline was more of an art than a science, relying on past experience. Several foundation-related engineering problems, such as the Leaning Tower of Pisa, prompted scientists to begin taking a more scientific-based approach to examining the subsurface. The application of the principles of mechanics to soils was documented as early as 1773 when Charles Coulomb (a physicist, engineer, and army Captain) developed improved methods to determine the earth pressures against military ramparts. Coulomb observed that, at failure, a distinct slip plane would form behind a sliding retaining wall and he suggested that the maximum shear stress on the slip plane, for design purposes, was the sum of the soil cohesion.

In the 19th century Henry Darcy developed what is now known as Darcy's Law describing the flow of fluids in porous media. Joseph Boussinesq (a mathematician and physicist) developed theories of stress distribution in elastic solids that proved useful for estimating stresses at depth in the ground. William Rankine, an engineer and physicist, developed an alternative to Coulomb's earth pressure theory. Albert Atterberg developed the clay consistency indices that are still used today for soil classification. Osborne Reynolds recognized in 1885 that shearing causes volumetric dilation of dense and contraction of loose granular materials.

Modern geotechnical engineering is said to have begun in 1925 with the publication of *Erdbaumechanik* by Karl Terzaghi (a civil engineer and geologist). Considered by many to be the father of modern soil mechanics and geotechnical engineering, Terzaghi developed the principle of effective stress, and demonstrated that the shear strength of soil is controlled by effective stress. Terzaghi also developed the framework for theories of bearing capacity of foundations, and the theory for prediction of the rate of settlement of clay layers due to consolidation. In his 1948 book, Donald Taylor recognized that interlocking and dilation of densely packed particles contributed to the peak strength of a soil. The interrelationships between volume change behavior (dilation, contraction, and consolidation) and shearing behavior were all connected via the theory of plasticity using critical state soil mechanics by Roscoe, Schofield, and Wroth with the publication of "On the Yielding of Soils" in 1958. Critical state soil mechanics is the basis for many contemporary advanced constitutive models describing the behavior of soil.

Geotechnical centrifuge modeling is a method of testing physical scale models of geotechnical problems. The use of a centrifuge enhances the similarity of the scale model tests involving soil because the strength and stiffness of soil is very sensitive to the confining pressure. The centrifugal acceleration allows a researcher to obtain large (prototype-scale) stresses in small physical models.

6.2 Foundation

A foundation (or, more commonly, base) is the element of an architectural structure which connects it to the ground, and transfers loads from the structure to the ground. Foundations are generally considered either shallow or deep. Foundation engineering is the application of soil mechanics and rock mechanics (geotechnical engineering) in the design of foundation elements of structures.[2]

6.2.1 Earthfast or post in ground construction

Buildings and structures have a long history of being built with wood in contact with the ground. Post in ground construction may technically have no foundation. Timber pilings were used on soft or wet ground even below stone or masonry walls. In marine construction and bridge building a crisscross of timbers or steel beams in concrete is called grillage.

Perhaps the simplest foundation is the padstone, a single stone which both spreads the weight on the ground and raises the timber off the ground. Staddle stones are a specific type of padstone.

Dry stone and stones laid in mortar to build foundations are common in many parts of the world. Dry laid stone foundations may have been painted with mortar after construction. Sometimes the top, visible course of stone is hewn, quarried stones. Besides using mortar, stones can also be put in a gabion. One disadvantage is that if using regular steel rebars, the gabion would last much less long than when using mortar (due to rusting). Using weathering steel rebars could reduce this disadvantage somewhat.

Rubble trench foundations are a shallow trench filled with rubble or stones. These foundations extend below the frost line and may have a drain pipe which helps groundwater drain away.

6.2.2 Modern foundation types

1. Shallow foundations

A shallow foundation is a type of foundation which transfers building loads to the earth very near the surface, rather than to a subsurface layer or a range of depths as does a deep foundation. Shallow foundations include spread footing foundations, mat-slab foundations, slab-on-grade foundations, pad foundations, rubble trench foundations and earthbag foundations.[3]

(1) Spread footing foundations

A spread footing foundation, which is typical in residential building, has a wider bottom portion than the load-bearing foundation walls it supports. This wider part "spreads" the weight of the structure over more area for greater stability.[4]

The design and layout of spread footings is controlled by several factors, foremost of which is the weight (load) of the structure it will support, as well as penetration of soft near-surface layers, and penetration through near-surface layers likely to change volume due to frost heave or shrink-swell.

These foundations are common in residential construction that includes a basement, and in many commercial structures. But for high rise buildings they are not sufficient.

A spread footing which changes elevation in several places in a series of vertical "steps" in order to follow the contours of a sloping site or accommodate changes in soil strata, is termed a stepped footing.

(2) Mat-slab foundations

Mat-slab foundations are used to distribute heavy column and wall loads across the entire building area, to lower the contact pressure compared to conventional spread footings. Mat-slab foundations can be constructed near the ground surface, or at the bottom of basements. In high-rise buildings, mat-slab foundations can be several meters thick, with extensive reinforcing to ensure relatively uniform load transfer.

(3) Slab-on-grade foundations

Slab-on-grade or floating slab foundations, as shown in Figure 6-1, are a structural engineering practice whereby the concrete slab that is to serve as the foundation for the structure is formed from a mold set into the ground. The concrete is then placed into the mold, leaving no space between the ground and the structure. This type of construction is most often seen in warmer climates, where ground freezing and thawing is less of a concern and where there is no need for heat ducting underneath the floor.

Figure 6-1　Example of Slab on Grade Foundation

The advantages of the slab technique are that it is cheap and sturdy, and is considered less vulnerable to termite infestation because there are no hollow spaces or wood channels leading from the ground to the structure (assuming wood siding, etc., is not carried all the way to the ground on the outer walls).

The disadvantages are the lack of access from below for utility lines, the potential for large heat losses where ground temperatures fall significantly below the interior temperature, and a very low elevation that exposes the building to flood damage in even moderate rains. Remodeling or extending such a structure may also be more difficult. Over the long term, ground settling (or subsidence) may be a problem, as a slab foundation cannot be readily jacked up to compensate; proper soil compaction prior to pour can minimize this. The slab can be decoupled from ground temperatures by insulation, with the concrete poured directly over insulation (for example, extruded polystyrene foam panels), or heating provisions (such as hydronic heating) can be built into the slab (an expensive installation, with associated running expenses).

Slab-on-grade foundations are commonly used in areas with expansive clay soil. While elevated structural slabs actually perform better on expansive clays, it is generally accepted by the engineering community that slab-on-grade foundations offer the greatest cost-to-performance ratio for tract homes. Elevated structural slabs are generally only found on custom homes or homes with basements.

Copper piping, commonly used to carry natural gas and water, reacts with concrete over a long period, slowly degrading until the pipe fails. This can lead to what is commonly referred to as slab leaks. These occur when pipes begin to leak from within the slab. Signs of a slab leak range from unexplained dampened carpet spots, to drops in water pressure and wet discoloration on exterior foundation walls. Copper pipes must be lagged (that is, insulated) or run through a conduit or plumbed into the building above the slab. Electrical conduits through the slab need to be water-tight, as they extend below ground level and can potentially expose the wiring to groundwater.

(4) Rubble trench foundations

The rubble trench foundation, as shown in Figure 6-2, a construction approach popularized by architect Frank Lloyd Wright, is a type of foundation that uses loose stone or rubble to minimize the use of concrete and improve drainage. It is considered more environmentally friendly than other types of foundation because cement manufacturing requires the use of enormous amounts of energy. However, some soil environments [such as particularly expansive or poor load-bearing (<1 ton/sf) soils] are not suitable for this kind of foundation.

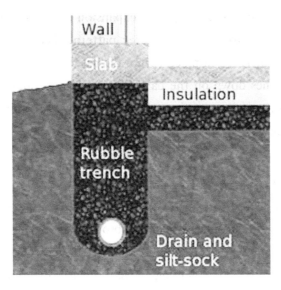

Figure 6-2 A Cross Section View of a Rubble Trench Foundation

A foundation must bear the structural loads imposed upon it and allow proper drainage of ground water to prevent expansion or weakening of soils and frost heaving. While the far more common concrete foundation requires separate measures to ensure good soil drainage, the rubble trench foundation serves both foundation functions at once.

To construct a rubble trench foundation a narrow trench is dug down below the frost line. The bottom of the trench would ideally be gently sloped to an outlet. Drainage tile is then placed at the bottom of the trench in a bed of washed stone protected by filter fabric. The trench is then filled with either screened stone or recycled rubble.

If an insulated slab is to be poured inside the grade beam, then the outer surface of the grade beam and the rubble trench should be insulated with rigid XPS foam board, which must be protected above grade from mechanical and UV degradation.

The rubble-trench foundation is a relatively simple, low-cost, and environmentally-friendly alternative to a conventional foundation, but may require an engineer's approval if building officials are not familiar with it. Frank Lloyd Wright used them successfully for more than 50 years in the first half of the 20th century, and there is a revival of this style of foundation with the increased interest in green building.

(5) Earthbag foundations

The basic construction method begins by digging a trench down to undisturbed mineral subsoil. Rows of woven bags (or tubes) are filled with available material, placed into this trench, compacted with a pounder to around 1/3 thickness of pre-pounded thickness, and form a foundation. Each successive layer will have one or more strands of barbed wire placed on top. This digs into the bag's weave and prevents slippage of subsequent layers, and also resists any tendency for the outward expansion of walls. The next row of bags is offset by half a bag's width to form a staggered pattern. These are either pre-filled with material and delivered, or filled in place (often the case with Superadobe). The weight of this earth-filled bag pushes down on the barbed wire strands, locking the bag in place on the row below. The same process continues layer upon layer, forming walls. A roof can be formed by gradually sloping the walls inward to construct a dome. Traditional types of roof can also be made.

2. Deep foundations

A deep foundation is used to transfer the load of a structure down through the upper weak layer of topsoil to the stronger layer of subsoil below. There are different types of deep footings including impact driven piles, drilled shafts, caissons, helical piles, geo-piers and earth stabilized columns.[5] The naming conventions for different types of footings vary between different engineers. Historically, piles were wood, later steel, reinforced concrete, and pre-tensioned concrete. A pile is a vertical structural element of a deep foundation, driven or drilled deep into the ground at the building site.

There are many reasons a geotechnical engineer would recommend a deep foundation over a shallow foundation, but some of the common reasons are very large design loads, a poor soil at shallow depth, or site constraints (like property lines). There are different terms used to describe different types of deep foundations including the pile (which is analogous to a pole), the pier (which is analogous to a column), drilled shafts, and caissons. Piles are generally driven into the ground in situ; other deep foundations are typically put in place using excavation and drilling. The naming conventions may vary between engineering disciplines and firms. Deep foundations (shown in Figure 6-3) can be made out of timber, steel, reinforced concrete or prestressed concrete.

Figure 6-3 A Deep Foundation Installation for a Bridge in Napa, California, United States

(1) Driven foundations

Prefabricated piles are driven into the ground using a pile driver. Driven piles are either wood, reinforced concrete, or steel. Wooden piles are made from the trunks of tall trees. Concrete piles are available in square, octagonal, and round cross-sections.[6] They are reinforced with rebar and are often prestressed. Steel piles are either pipe piles or some sort of beam section (like an H-pile). Historically, wood piles used splices to join multiple segments end-to-end when the driven depth required was too long for a single pile; today, splicing is common with steel piles, though concrete piles can be spliced with mechanical and other means. Driving piles, as opposed to drilling shafts, is advantageous because the soil displaced by driving the piles compresses the surrounding soil, causing greater friction against the sides of the piles, thus increasing their load-bearing capacity.

Foundations relying on driven piles often have groups of piles connected by a pile cap (a large concrete block into which the heads of the piles are embedded) to distribute loads which are larger than one pile can bear. Pile caps and isolated piles are typically connected with grade beams to tie the foundation elements together; lighter structural elements bear on the grade beams, while heavier elements bear directly on the pile cap.

(2) Drilled piles

Drilled piles, also called caissons, drilled shafts, drilled piers, Cast-in-drilled-hole piles (CIDH piles) or Cast-in-Situ piles, are largely used in the field of deep foundations. A borehole is drilled into the ground, then concrete (and often some sort of reinforcing) is placed into the borehole to form the pile. Rotary boring techniques allow larger diameter piles than any other piling method and permit pile construction through particularly dense or hard strata. Construction methods depend on the geology of the site; in particular, whether boring is to be undertaken in "dry" ground conditions or through water-saturated strata-i. e. "wet boring".[7] Casing is often used when the sides of the borehole are likely to slough off before concrete is poured.

For end-bearing piles, drilling continues until the borehole has extended a sufficient depth (socketing) into a sufficiently strong layer. Depending on site geology, this can be a rock layer, or hardpan, or other dense, strong layers. Both the diameter of the pile and the depth of the pile are highly specific to the ground conditions, loading conditions, and nature of the project. Pile depths may vary substantially across a project if the bearing layer is not level. Drilled piles can be tested using a variety of methods to verify the pile integrity during installation.

Glossary 词汇表

geotechnical [dʒiːəʊˈteknɪkəl] *adj*. 岩土工程的
earthwork [ˈɜːθwɜːk] *n*. 土方（工程）
rockfall [ˈrɒkfɔːl] *n*. 岩崩，落石
soil liquefaction 土壤液化
debris flow 泥石流
sinkhole [ˈsɪŋkhəʊl] *n*. 天坑
embankment [ɪmˈbæŋkmənt] *n*. 路堤，筑堤
tunnel [ˈtʌnl] *n*. 隧道
dike [daɪk] *n*. 堤
levee [ˈlevi] *n*. 堤，堤坝
channel [ˈtʃænl] *n*. 渠道
reservoir [ˈrezəvwɑː(r)] *n*. 蓄水池
sanitary [ˈsænətri] *adj*. 卫生的
landfill [ˈlændfɪl] *n*. 废渣填埋

wharves [wɔːvz] *n*. 码头，停泊处
marina [məˈriːnə] *n*. 小艇船坞
jetty [ˈdʒeti] *n*. 码头，防波堤，
　[建] 建筑物的突出部
centrifuge [ˈsentrɪfjuːdʒ] *vt*. 使离心
earthfast [ɜːθˈfɑːst] *adj*. 牢固在土中的
padstone [ˈpædstəʊn] *n*. 垫石
hewn [hjuːn] *adj*. 开凿的
trench [trentʃ] *n*. 沟，渠
shallow foundation 浅基础

spread footing foundation 扩展基础
mat-slab foundation 片筏基础
slab-on-grade foundation 平板式基础
settling [ˈsetlɪŋ] *n*. 固结，沉降
subsidence [səbˈsaɪdns] *n*. 沉降
rubble trench foundation 碎石沟基础
earthbag foundation 土工袋基础
deep foundation 深基础
driven foundation 夯实基础
drilled pile 钻孔桩

Notes 注释

[1] Geotechnical engineering is the branch of civil engineering concerned with the engineering behavior of earth materials. Geotechnical engineering is important in civil engineering, but also has applications in military, mining, petroleum and other engineering disciplines that are concerned with construction occurring on the surface or within the ground.

　　岩土工程是土木工程中的一个分支，它涉及土体材料的工程特性。岩土工程在土木工程中占有重要地位，在军事、采矿、石油等涉及地表或地下施工的工程学科中也有着广泛的应用。

[2] A foundation (or, more commonly, base) is the element of an architectural structure which connects it to the ground, and transfers loads from the structure to the ground. Foundations are generally considered either shallow or deep. Foundation engineering is the application of soil mechanics and rock mechanics (Geotechnical engineering) in the design of foundation elements of structures.

　　基础是建筑结构的一部分，它连接到地基，并将荷载从结构传递到地基。基础通常可分为浅基础或深基础。基础工程学是土力学和岩石力学（岩土工程）在结构基础单元设计中的应用。

[3] A shallow foundation is a type of foundation which transfers building loads to the earth very near the surface, rather than to a subsurface layer or a range of depths as does a deep foundation. Shallow foundations include spread footing foundations, mat-slab foundations, slab-on-grade foundations, pad foundations, rubble trench foundations and earthbag foundations.

浅基础将建筑荷载转移到地表附近，而不像深基础那样将其转移到地下层或一段深度。浅基础包括扩展基础、片筏基础、平板式基础、独立基础、碎石沟基础和土工袋基础。

[4] A spread footing foundation, which is typical in residential building, has a wider bottom portion than the load-bearing foundation walls it supports. This wider part "spreads" the weight of the structure over more area for greater stability.

扩展基础在住宅建筑很典型，其底部比其上部所支撑的承重基础墙更宽。这个更广泛的部分将结构的重量"传递"到更广的范围，从而获得更大的稳定性。

[5] A deep foundation is used to transfer the load of a structure down through the upper weak layer of topsoil to the stronger layer of subsoil below. There are different types of deep footings including impact driven piles, drilled shafts, caissons, helical piles, geo-piers and earth stabilized columns.

深基础用于将结构的载荷通过表层的上层软弱土层传递到更强的下层土层。不同类型的深基础包括冲击桩、钻孔桩、沉箱、螺旋桩、墩柱和土体稳定柱。

[6] Prefabricated piles are driven into the ground using a pile driver. Driven piles are either wood, reinforced concrete, or steel. Wooden piles are made from the trunks of tall trees. Concrete piles are available in square, octagonal, and round cross-sections.

预制桩用打桩机打入地面。打入桩可以是木桩、钢筋混凝土桩或钢桩。木桩是由树干或高大树木做成的。混凝土桩有方形、八角形和圆形截面。

[7] A borehole is drilled into the ground, then concrete (and often some sort of reinforcing) is placed into the borehole to form the pile. Rotary boring techniques allow larger diameter piles than any other piling method and permit pile construction through particularly dense or hard strata. Construction methods depend on the geology of the site; in particular, whether boring is to be undertaken in "dry" ground conditions or through water-saturated strata - i. e. "wet boring".

将钻孔钻入地下，然后将混凝土（通常是某种加固件）放入钻孔中以形成桩。旋转钻孔技术允许采用比任何其他打桩方法更大直径的桩，并允许通过特别密集或坚硬的地层进行桩施工。施工方法取决于场地的地质情况，特别要注意钻孔是在"干燥的"地质条件下或者是通过含水饱和地层（即湿法成孔）进行的。

References 课文来源

[1] https://en. wikipedia. org/wiki/Shallow _ foundation
[2] https://en. wikipedia. org/wiki/Rubble _ trench _ foundation

Unit 7 Surveying

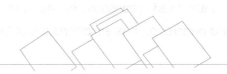

- 学习任务 测量工程
- 教学时间 2 学时
- 学习目标 了解测量技术的相关术语，重点掌握测量设备的技术特点。

Text 课文

Surveying or land surveying is the technique, profession, and science of determining the terrestrial or three-dimensional position of points and the distances and angles between them. A land surveying professional is called a land surveyor (shown in Figure 7-1). These points are usually on the surface of the earth, and they are often used to establish land maps and boundaries for ownership, locations like building corners or the surface location of subsurface features, or other purposes required by government or civil law, such as property sales.

Figure 7-1 A Surveyor at Work with an Infrared Reflector Used for Distance Measurement

Surveyors work with elements of geometry, trigonometry, regression analysis, physics, engineering, metrology, programming languages and the law. They use equipment like total stations, robotic total stations, GPS receivers, retroreflectors, 3D scanners, radios, handheld tablets, digital levels, subsurface locators, drones, GIS and surveying software.[1]

Surveying has been an element in the development of the human environment since the beginning of recorded history. The planning and execution of most forms of construction require it. It is also used in transport, communications, mapping, and the definition of legal boundaries for land ownership. It is an important tool for research in many other scientific disciplines.

7.1 History

7.1.1 Ancient surveying

Basic surveyance has occurred since humans built the first large structures. In ancient Egypt, a rope stretcher would use simple geometry to re-establish boundaries after the annual floods of the Nile River. The almost perfect squareness and north-south orientation of the Great Pyramid of Giza, built c. 2700 BC, affirm the Egyptians' command of surveying. The Groma instrument originated in Mesopotamia (early 1st millennium BC). The mathematician Liu Hui described ways of measuring distant objects in his work *Haidao Suanjing* or *The Sea Island Mathematical Manual*, published in 263 AD. The Romans recognized land surveyors as a profession. They established the basic measurements under which the Roman Empire was divided, such as a tax register of conquered lands (300 AD). Roman surveyors were known as Gromatici.

In medieval Europe, beating the bounds maintained the boundaries of a village or parish. This was the practice of gathering a group of residents and walking around the parish or village to establish a communal memory of the boundaries. Young boys were included to ensure the memory lasted as long as possible. In England, William the Conqueror commissioned the *Domesday Book* in 1086. It recorded the names of all the land owners, the area of land they owned, the quality of the land, and specific information of the area's content and inhabitants. It did not include maps showing exact locations.

7.1.2 Modern surveying

In the 18th century, modern techniques and instruments for surveying began to be used. Jesse Ramsden introduced the first precision theodolite in 1787. It was an instrument for measuring angles in the horizontal and vertical planes. He created his great theodolite using an accurate dividing engine of his own design. Ramsden's theodolite represented a great step forward in the instrument's accuracy. William Gascoigne invented an instrument that used a telescope with an installed crosshair as a target device, in 1640. James Watt developed an optical meter for the measuring of distance in 1771; it measured the parallactic angle from which the distance to a point could be deduced.

Dutch mathematician Willebrord Snellius (a. k. a. Snel van Royen) introduced the modern systematic use of triangulation. In 1615 he surveyed the distance from Alkmaar to Breda, approximately 72 miles (116,1 kilometres). Between 1733 and 1740, Jacques Cassini and his son César undertook the first triangulation of France. They included a re-surveying of the meridian arc, leading to the publication in 1745 of the first map of France constructed on rigorous principles. By this time, triangulation methods were by then well established for local map-making.

It was only towards the end of the 18th century that detailed triangulation network surveys mapped whole countries. In 1784, a team from General William Roy's Ordnance Survey of Great Britain began the Principal Triangulation of Britain.

In the US, the *Land Ordinance* of 1785 created the Public Land Survey System. It formed the basis for dividing the western territories into sections to allow the sale of land. The PLSS divided states into township grids which were further divided into sections and fractions of sections.

7.1.3　Recent surveying

At the beginning of the century surveyors had improved the older chains and ropes, but still faced the problem of accurate measurement of long distances. Dr Trevor Lloyd Wadley developed the Tellurometer during the 1950s. It measures long distances using two microwave transmitter/receivers. During the late 1950s Geodimeter introduced electronic distance measurement (EDM) equipment.

In the 1970s the first instruments combining angle and distance measurement appeared, becoming known as total stations. Manufacturers added more equipment by degrees, bringing improvements in accuracy and speed of measurement. Major advances include tilt compensators, data recorders, and on-board calculation programs.

The first satellite positioning system was the US Navy TRANSIT system. The first successful launch took place in 1960. The system's main purpose was to provide position information to Polaris missile submarines. Surveyors found they could use field receivers to determine the location of a point. Sparse satellite cover and large equipment made observations laborious, and inaccurate. The main use was establishing benchmarks in remote locations.

The US Air Force launched the first prototype satellites of the Global Positioning System (GPS) in 1978. GPS used a larger constellation of satellites and improved signal transmission to provide more accuracy. Early GPS observations required several hours of observations by a static receiver to reach survey accuracy requirements. Recent improvements to both satellites and receivers allow Real Time Kinematic (RTK) surveying.

7.1.4 Current surveying

Remote sensing and satellite imagery continue to improve and become cheaper, allowing more commonplace use. Prominent new technologies include three-dimensional (3D) scanning and use of lidar for topographical surveys. UAV technology along with photogrammetric image processing is also appearing.

7.2 Surveying Equipment

The main surveying instruments (shown in Figure 7-2) in use around the world are the theodolite, measuring tape, total station, 3D scanners, GPS/GNSS, level and rod. Most instruments screw onto a tripod when in use. Tape measures are often used for measurement of smaller distances. 3D scanners and various forms of aerial imagery are also used.[2]

Figure 7-2　Surveying Equipment

a) optical theodolite　b) robotic total station　c) RTK GPS base station　d) optical level

The theodolite is an instrument for the measurement of angles. It uses two separate circles, protractors or alidades to measure angles in the horizontal and the vertical plane. A telescope mounted on trunnions is aligned vertically with the target object. The whole upper section rotates for horizontal alignment. The vertical circle measures the angle that the telescope makes against the vertical, known as the zenith angle. The horizontal circle uses an upper and lower plate.[3] When beginning the survey, the surveyor points the instrument in a known direction (bearing), and clamps the lower plate in place. The instrument can then rotate to measure the bearing to other objects. If no bearing is known or direct angle measurement is wanted, the instrument can be set to zero during the initial sight. It will then read the angle between the initial object, the theodolite itself, and the item that the telescope aligns with.

The gyrotheodolite is a form of theodolite that uses a gyroscope to orient itself in the absence of reference marks. It is used in underground applications.

The total station is a development of the theodolite with an electronic distance measurement device (EDM). A total station can be used for leveling when set to the horizontal plane. Since their introduction, total stations have shifted from optical-mechanical to fully electronic devices.

Modern top-of-the-line total stations no longer need a reflector or prism to return the light pulses used for distance measurements. They are fully robotic, and can even e-mail point data to a remote computer and connect to satellite positioning systems, such as Global Positioning System. Real time kinematic GPS systems have increased the speed of surveying, but they are still only horizontally accurate to about 20 mm and vertically to 30 – 40 mm.

GPS surveying differs from other GPS uses in the equipment and methods used. Static GPS uses two receivers placed in position for a considerable length of time. The long span of time lets the receiver compare measurements as the satellites orbit.

RTK surveying uses one static antenna and one roving antenna. The static antenna tracks changes in the satellite positions and atmospheric conditions. The surveyor uses the roving antenna to measure the points needed for the survey. The two antennas use a radio link that allows the static antenna to send corrections to the roving antenna. The roving antenna then applies those corrections to the GPS signals it is receiving to calculate its own position.

Surveying instruments have characteristics that make them suitable for certain uses. Theodolites and levels are often used by constructors rather than surveyors in first world countries. The constructor can perform simple survey tasks using a relatively cheap instrument. Total stations are workhorses for many professional surveyors because they are versatile and reliable in all conditions. The productivity improvements from a GPS on large scale surveys makes them popular for major infrastructure or data gathering projects. One-person robotic-guided total stations allow surveyors to measure without extra workers to aim the telescope or record data. A fast but expensive way to measure large areas is with a helicopter, using a GPS to record the location of the helicopter and a laser scanner to measure the ground. To increase precision, surveyors place beacons on the ground [about 20 km (12 mi) apart]. This method reaches precisions between 5 – 40 cm (depending on flight height).

7.3 Surveying Techniques

Surveyors determine the position of objects by measuring angles and distances. The factors that can affect the accuracy of their observations are also measured. They then use this data to create vectors, bearings, coordinates, elevations, areas, volumes, plans and maps. Measurements are often split into horizontal and vertical components to simplify calculation. GPS and astronomic measurements also need measurement of a time component.[4]

7.3.1 Distance measurement

Before EDM devices, distances were measured using a variety of means. These included chains with links of a known length such as a Gunter's chain, or measuring tapes made of steel or invar. To measure horizontal distances, these chains or tapes were pulled taut to reduce sagging and slack. The distance had to be adjusted for heat expansion. Attempts to hold the measuring instrument level would also be made. When measuring up a slope, the surveyor might have to "break" (break chain) the measurement—use an increment less than the total length of the chain. Perambulators, or measuring wheels, were used to measure longer distances but not to a high level of accuracy. Tacheometry is the science of measuring distances by measuring the angle between two ends of an object with a known size. It was sometimes used before to the invention of EDM where rough ground made chain measurement impractical.

7.3.2 Angle measurement

Historically, horizontal angles were measured by using a compass to provide a magnetic bearing or azimuth. Later, more precise scribed discs improved angular resolution. Mounting telescopes with reticles atop the disc allowed more precise sighting (see theodolite). Levels and calibrated circles allowed measurement of vertical angles.

By observing the bearing from every vertex in a figure, a surveyor can measure around the figure. The final observation will be between the two points first observed, except with a 180° difference. This is called a close. If the first and last bearings are different, this shows the error in the survey, called the angular misclose. The surveyor can use this information to prove that the work meets the expected standards.

7.3.3 Levelling

The simplest method for measuring height is with an altimeter using air pressure to find height. When more precise measurements are needed, means like precise levels (also known as differential leveling) are used. When precise leveling, a series of measurements between two points are taken using an instrument and a measuring rod. Differences in height between the measurements are added and subtracted in a series to get the net difference in elevation between the two endpoints. With the Global Positioning System (GPS), elevation can be measured with satellite receivers. Usually GPS is somewhat less accurate than traditional precise leveling, but may be similar over long distances.[5]

When using an optical level, the endpoint may be out of the effective range of the instrument. There may be obstructions or large changes of elevation between the endpoints. In these situations, extra setups are needed. Turning is a term used when referring to moving the level to take an elevation shot from a different location. To "turn" the level, one must first take a reading and record the elevation of the point the rod is located on. While the rod is being kept in exactly the same location, the level is moved to a new location where the rod is still visible. A reading is taken from the new location of the level and the height difference is used to find the new elevation of the level gun. This is repeated until the series of measurements is completed. The level must be horizontal to get a valid measurement. Because of this, if the horizontal crosshair of the instrument is lower than the base of the rod, the surveyor will not be able to sight the rod and get a reading. The rod can usually be raised up to 25 feet high, allowing the level to be set much higher than the base of the rod.

7.3.4 Errors and accuracy

A basic tenet of surveying is that no measurement is perfect, and that there will always be a small amount of error. There are three classes of survey errors.

(1) Gross errors or blunders: Errors made by the surveyor during the survey. Upsetting the instrument, misaiming a target, or writing down a wrong measurement are all gross errors. A large gross error may reduce the accuracy to an unacceptable level. Therefore, surveyors use redundant measurements and independent checks to detect these errors early in the survey.

(2) Systematic errors: Systematic errors are caused by imperfect calibration of measurement instruments or imperfect methods of observation, or interference of the

environment with the measurement process, and always affect the results of an experiment in a predictable direction. Incorrect zeroing of an instrument leading to a zero error is an example of systematic error in instrumentation.

(3) Random errors: Random errors are small unavoidable fluctuations. They are caused by imperfections in measuring equipment, eyesight, and conditions. They can be minimized by redundancy of measurement and avoiding unstable conditions. Random errors tend to cancel each other out, but checks must be made to ensure they are not propagating from one measurement to the next.[6]

Surveyors avoid these errors by calibrating their equipment, using consistent methods, and by good design of their reference network. Repeated measurements can be averaged and any outlier measurements discarded. Independent checks like measuring a point from two or more locations or using two different methods are used. Errors can be detected by comparing the results of the two measurements.

Once the surveyor has calculated the level of the errors in his work, it is adjusted. This is the process of distributing the error between all measurements. Each observation is weighted according to how much of the total error it is likely to have caused and part of that error is allocated to it in a proportional way. The most common methods of adjustment are the Bowditch method, also known as the compass rule, and the Principle of least squares method.

Glossary 词汇表

surveying [sɜː'veɪɪŋ] n. 测量学
trigonometry [ˌtrɪgə'nɒmətri] n. 三角学
regression analysis 回归分析
metrology [mə'trɒlədʒɪ] n. 计量学
digital level 数字水准仪
subsurface locator 地下定位器
drone [drəʊn] n. 无人驾驶飞机
prototype ['prəʊtətaɪp] n. 原型
constellation [ˌkɒnstə'leɪʃn] n. 星座
lidar ['laɪdə] n. 激光雷达
topographical [ˌtɒpə'græfɪkl] adj. 地形学的
photogrammetric [ˌfəʊtəʊgrə'metrɪk] adj. 摄影测量的

theodolite [θi'ɒdəlaɪt] n. 经纬仪
robotic total station 智能型全站仪
protractor [prə'træktə(r)] n. 量角器
alidade [ˌæliː'deɪd] n. 照准仪
zenith ['zenɪθ] n. 顶点
clamp [klæmp] vt. & vi. 夹紧
gyrotheodolite [dʒaɪ'rɒθiːəʊdəlaɪt] n. 陀螺经纬仪
gyroscope ['dʒaɪrəskəʊp] n. 陀螺仪，回转仪
Global Positioning System 全球定位系统
coordinate [kəʊ'ɔːdɪneɪt] n. 坐标
elevation [ˌelɪ'veɪʃn] n. 高程

perambulator [pəˈræmbjuleɪtə(r)] n. 测距仪
tacheometry [tækɪˈɒmɪtrɪ] n. 视距测量法
azimuth [ˈæzɪməθ] n. 方位角
misclose n. 闭合差
levelling [ˈlevəlɪŋ] n. 抄平
endpoint [ˈendpɔɪt] n. 端点
gross error 粗差
blunder [ˈblʌndə(r)] n. 粗差
least squares method 最小二乘法

Notes 注释

[1] Surveyors work with elements of geometry, trigonometry, regression analysis, physics, engineering, metrology, programming languages and the law. They use equipment like total stations, robotic total stations, GPS receivers, retroreflectors, 3D scanners, radios, handheld tablets, digital levels, subsurface locators, drones, GIS and surveying software.

测量员使用几何学、三角学、回归分析、物理学、工程学、计量学、程序设计语言和法律等要素进行工作。他们使用的设备有如全站仪、自动全站仪、GPS 接收机、后向反射器、3D 扫描仪、无线电、手持平板电脑、数字水准仪、地下定位器、无人机、GIS 和测量软件。

[2] The main surveying instruments (shown in Figure 7-2) in use around the world are the theodolite, measuring tape, total station, 3D scanners, GPS/GNSS, level and rod. Most instruments screw onto a tripod when in use. Tape measures are often used for measurement of smaller distances. 3D scanners and various forms of aerial imagery are also used.

世界各地使用的主要测量仪器有经纬仪、卷尺、全站仪、3D 扫描仪、GPS／GNSS、水准仪和标尺（图 7-2）。大多数仪器在使用时都会拧到三脚架上。卷尺常用来测量较小的距离。3D 扫描仪和各种形式的航空图像也会被使用。

[3] The theodolite is an instrument for the measurement of angles. It uses two separate circles, protractors or alidades to measure angles in the horizontal and the vertical plane. A telescope mounted on trunnions is aligned vertically with the target object. The whole upper section rotates for horizontal alignment. The vertical circle measures the angle that the telescope makes against the vertical, known as the zenith angle. The horizontal circle uses an upper and lower plate.

经纬仪是一种测量角度的仪器。它使用了两个独立的圆圈，量角器或照准仪来测量水平和垂直平面的角度。安装在耳轴上的望远镜与目标对象垂直对齐。整个上部旋转以进行水平对齐。垂直圆测量望远镜相对于垂直方向的角度，称为天顶角。水平圆采用上下板。

［4］ Surveyors determine the position of objects by measuring angles and distances. The factors that can affect the accuracy of their observations are also measured. They then use this data to create vectors, bearings, coordinates, elevations, areas, volumes, plans and maps. Measurements are often split into horizontal and vertical components to simplify calculation. GPS and astronomic measurements also need measurement of a time component.

测量师通过测量角度和距离来确定物体的位置，也会测量其他影响观测精度的因素。然后，测量员使用这些数据来创建矢量、方位、坐标、高程、面积、体积、计划和地图。测量通常被分解为水平和垂直分量，以简化计算。GPS 和天文测量还需要一个时间分量的测量。

［5］ The simplest method for measuring height is with an altimeter using air pressure to find height. When more precise measurements are needed, means like precise levels (also known as differential leveling) are used. When precise leveling, a series of measurements between two points are taken using an instrument and a measuring rod. Differences in height between the measurements are added and subtracted in a series to get the net difference in elevation between the two endpoints. With the Global Positioning System (GPS), elevation can be measured with satellite receivers. Usually GPS is somewhat less accurate than traditional precise leveling, but may be similar over long distances.

测量高度的最简单的方法是使用高度计利用空气压力测量高度。当需要更精确的测量时，使用诸如精确水平（也称为差分水准）的方法。进行精密水准测量时，使用仪器和测量尺进行两点之间的一系列测量。测量的高度之间的差异通过级数相加和相减得到两个端点之间的净高差。借助于全球定位系统（GPS），卫星接收机也可以测量高程。通常 GPS 比传统的精密水准测量精度要低一些，但在长距离上精度可能相似。

［6］ There are three classes of survey errors：

（1）Gross errors or blunders：Errors made by the surveyor during the survey. Upsetting the instrument, misaiming a target, or writing down a wrong measurement are all gross errors. A large gross error may reduce the accuracy to an unacceptable level. Therefore, surveyors use redundant measurements and independent checks to detect these errors early in the survey.

（2）Systematic errors：Systematic errors are caused by imperfect calibration of measurement instruments or imperfect methods of observation, or interference of the environment with the measurement process, and always affect the results of an experiment in a predictable direction. Incorrect zeroing of an instrument leading to a zero error is an example of systematic error in instrumentation.

(3) Random errors: Random errors are small unavoidable fluctuations. They are caused by imperfections in measuring equipment, eyesight, and conditions. They can be minimized by redundancy of measurement and avoiding unstable conditions. Random errors tend to cancel each other out, but checks must be made to ensure they are not propagating from one measurement to the next.

测量误差有三类:

(1) 粗差或误差: 测量员在测量过程中发生失误。如颠倒仪器、目标失误或错误测量记录都属于粗差。大的粗差可能会使精度降低到不可接受的水平。因此,测量员在测量早期使用冗余测量和独立检查来检测这些错误。

(2) 系统误差: 系统误差是由于测量仪器的不完美校准或不完善的观测方法,或测量过程对环境的干扰引起的,并且总是以可预测的方向影响测量结果。测量仪器的不正确归零所导致零点误差是系统误差在测量仪器上的一个例子。

(3) 随机误差: 随机误差是不可避免的小的波动。它们是由于测量设备、视力和条件的不完善所引起的。随机误差可以通过冗余测量和避免不稳定的条件来最小化。随机误差倾向于彼此抵消,但必须进行检查以确保它们不会从一个测量值传递给下一个测量值。

References 课文来源

[1] http://hitchhikersgui.de/Surveying
[2] https://en.wikipedia.org/wiki/Surveying

Unit 8　Building Facade Types

> ▶ 学习任务　　幕墙
> ▶ 教学时间　　2 学时
> ▶ 学习目标　　了解幕墙结构的相关术语，了解幕墙的节点体系构成。

Text 课文

New wall and cladding systems are tailored to meet the aesthetic and performance requirements of a particular building. The facade types include unitized curtain walls, unitized skylights, point-fixed glass walls, lightweight truss-supported glazing, and also claddings. [1]

8.1　Curtain Wall Design

A curtain wall system, as shown in Figure 8-1, is an outer covering of a building in which the outer walls are non-structural, but merely keep the weather out and the occupants in. As the curtain wall is non-structural, it can be made of a lightweight material, reducing construction costs. When glass is used as the curtain wall, a great advantage is that natural light can penetrate deeper within the building. [2] The curtain wall façade does not carry any dead load weight from the building other than its own dead load weight. The wall transfers horizontal wind loads that are incident upon it to the main building structure through connections at floors or columns of the building. A curtain wall is designed to resist air and water infiltration absorb, sway induced by wind and seismic forces acting on the building, and support its own dead load weight forces.

Curtain-wall systems are typically designed with extruded aluminum members, although the first curtain walls were made of steel. The aluminium frame is typically infilled with glass, which provides an architecturally pleasing building, as well as benefits such as daylighting. However, parameters related to solar gain control such as thermal comfort and visual comfort are more difficult to control when using highly glazed curtain walls. Other common infills include: stone veneer, metal panels, louvres, and operable windows or vents.

Figure 8-1 Aluminium Thermal Break Curtain Wall and Customized Invisible Frame Curtain Wall

Curtain walls differ from store-front systems in that they are designed to span multiple floors, and take into consideration design requirements such as: thermal expansion and contraction; building sway and movement; water diversion; and thermal efficiency for cost-effective heating, cooling, and lighting in the building.

8.2 Cladding

Cladding, as shown in Figure 8-2, is the application of one material over another to provide a skin or layer. In construction, cladding is used to provide a degree of thermal insulation and weather resistance, and to improve the appearance of buildings. Cladding can be made of any of a wide range of materials including wood, metal, brick, vinyl, and composite materials that can include aluminium, wood, blends of cement and recycled polystyrene, wheat/rice straw fibres.[3] Rainscreen cladding is a form of weather cladding designed to protect against the elements, but also offers thermal insulation. The cladding does not itself need to be waterproof, merely a control element: it may serve only to direct water or wind safely away in order to control run-off and prevent its infiltration into the building structure. Cladding may also be a control element for noise, either entering or escaping. Cladding can become a fire risk by design or material. The types of facing with which we are most familiar are solid aluminium sheet, composite panel, natural stone and terracotta: we are however always interested to work on projects with adventurous architects who wish to use more exotic materials.

Figure 8-2　Cladding with Geometrically Complex Features

8.3　Lightweight Rod and Cable Truss Systems

Within the public areas of a grand building there may be an architectural requirement for glazing that is high or wide in span, but unobstructed by conventional structural beams or columns. The lightweight truss supports (shown in Figure 8-3), made up of rod or cable elements, is design to satisfy this requirement.

Figure 8-3　Lightweight Truss Supports
a) a vertical cable truss capable of resisting typhoon storm conditions
b) a horizontal rod truss at an aluminium-framed curtain wall

8.4 Skylights

A number of different "breeds" of skylight system (shown in Figure 8-4), custom-designed extrusions for use overhead in glass roofs or skylights, have evolved in design office: the simplest are extruded veneers that attach to steel members; other types use only aluminium framings, and can be either stick built at site or panelized to allow glazing operations to be carried out in a fabrication workshop. The internal gutters are incorporated to collect rain water from the purlins and from the rafters, preventing leaks to the interior of the building.

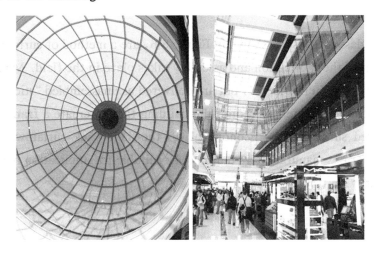

Figure 8-4 Skylights
a) a veneer-type glazing system applied to a dome-shaped steel shell structure
b) a stick-built, aluminium-framed skylight at an airport terminal

8.5 Structural Glass Facades

A new facade technology has gradually emerged in recent decades, driven largely by the pursuit of transparency in the building facade among leading international building designers. This new technology has evolved in long-span applications, and can be categorized by the various structural systems employed as support.

8.5.1 Framed

Framed systems, as shown in Figure 8-5, support the glass continuously along two or four sides. There are many variations of framed systems, most of which fall into two general categories. Conventional unitized curtainwall systems are seldom used with structural glass facades.

Figure 8-5　Framed System

8.5.2　Stick

Stick-built glass facades, as shown in Figure 8-6, are a method of curtainwall construction where much of the fabrication and assembly takes place in the field. Mullions of extruded aluminum may be prefabricated, but are delivered as unassembled "sticks" to the building site. Mullions are then installed onto the building face to create a frame for the glass, which is installed subsequently. Economical off-the-shelf stick curtainwall products are available from various manufacturers that may be suitable for application in structural glass facades, primarily on truss systems.

Figure 8-6　Stick System

8.5.3　Veneer

Truss systems can be designed with an outer chord of square or rectangular tubing, and may include transom components of similar material, presenting a uniform flat grid installed to high tolerances. Such a system can provide continuous support to the simplest and most minimal off-the-shelf glazing system, thus combining relatively high transparency with excellent economy. A veneer glazing system, as shown in Figure 8-7, is essentially a stick-built curtainwall system designed for continuous support and representing a higher level of system integration with resulting efficiencies. Variations can include 4-sided capture, 2-sided capture, structurally glazed and unitized systems.

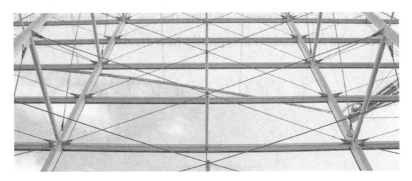

Figure 8-7　Stick-built Glass Facade

8.5.4　Panel / Cassette

Panel systems, as shown in Figure 8-8, are typically constructed from a framed glass lite. The framed panel can then be point-supported by a supporting structural system, while the glass remains continuously supported on two or four sides. This also allows the panel to be stepped away from the support system—a practice that visually lightens the facade. Panel systems can be prefabricated, benefiting from assembly under factory-controlled conditions.

Cassette systems combine properties of stick, veneer and panel systems. While variations exist, the predominant makeup of a cassette system is comprised of a primary structural mullion system, which is stick built. These provide the support and facilitate the attachment of the glass panels. The glass lites are factory assembled into minimal frames, which form an integral connection with the primary mullion system. A cassette system can be designed to be fully shop-glazed, requiring no application of sealant during field installation.

Figure 8-8　Panel / Cassette System

8.5.5 Frameless

Frameless systems, as shown in Figure 8-9, utilize glass panes that are fixed to a structural system at discrete points, usually near the corners of the glass panel (point-fixed). The glass is directly supported without the use of perimeter framing elements. Glass used in point-fixed applications is typically heat-treated.

Figure 8-9　Frameless System

8.5.6 Point-Fixed Bolted

The most popular (and often most expensive) glass system for application in structural glass facades is the bolted version (shown in Figure 8-10). The glass panel requires perforations to accommodate specialized bolting hardware. Specially designed off-the-shelf hardware systems are readily available, or custom components can be designed. [4] Cast stainless steel spider fittings are most commonly used to tie the glass to the supporting structure, although custom fittings are often developed for larger facade projects. The glass must be designed to accommodate bending loads and deflections resulting from the fixing method. For overhead applications, insulated-laminated glass panels require the fabrication of 12 holes per panel, which can represent a cost constraint on some projects.

Figure 8-10　Point-Fixed Bolted System

8.5.7 Point-Fixed Clamped

Point-fixed clamped systems, as shown in Figure 8-11, are a solution for point fixing without the perforations in glass. In the case of a spider type fitting, the spider is rotated 45 degrees from the bolted position so that its arms align with glass seams. A thin blade penetrates through the seam between adjacent pieces of glass. An exterior plate attaches to the blade and clamps the glass in place. The bolted systems present an uninterrupted glass surface, while the clamped systems expose the small exterior clamp plate.[5] Some facade designers prefer the exposed hardware aesthetic. While clamped systems have the potential for greater economy by eliminating the need for glass perforations, the cost of the clamping hardware may offset at least some savings, depending upon the efficiency of the design.

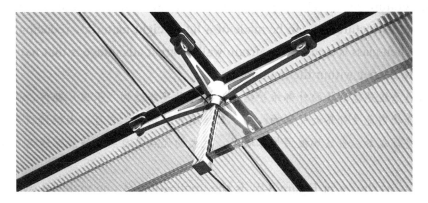

Figure 8-11 Point-Fixed Clamped System

Glossary 词汇表

facade [fə'sɑːd] n. 外表，建筑物的正面
cladding ['klædɪŋ] n. 覆层
aesthetic [iːs'θetɪk] adj. 审美的，美学的
curtain wall 幕墙
skylight ['skaɪlaɪt] n. 天窗
point-fixed glass wall 点固定玻璃墙
lightweight truss-supported glazing 轻型桁架支撑玻璃
infiltration [ˌɪnfɪl'treɪʃn] n. 渗透
aluminium [ˌæljə'mɪniəm] n. 铝

veneer [və'nɪə(r)] n. 饰面，外饰
louvre ['luːvə(r)] n. 百叶窗
operable window 活动窗
vent [vent] n. 通风口
glass lites n. 玻璃片
frameless ['freɪmlɪs] adj. 无框架的
perforation [ˌpɜːfə'reɪʃn] n. 穿孔，贯穿
seam [siːm] n. 接缝
point-fixed bolted system 点固定螺栓系统
point-fixed clamped system 点固定夹紧系统

Notes 注释

[1] New wall and cladding systems are tailored to meet the aesthetic and performance requirements of a particular building. The facade types include unitized curtain walls, unitized skylights, point-fixed glass walls, lightweight truss-supported glazing, and also claddings.

新的墙壁和覆层系统是为满足特定建筑物的美学和性能要求而量身定做的。立面类型包括组合式幕墙、组合式天窗、点固定玻璃墙、轻型桁架支撑玻璃以及覆层。

[2] A curtain wall system is an outer covering of a building in which the outer walls are non-structural, but merely keep the weather out and the occupants in. As the curtain wall is non-structural, it can be made of a lightweight material, reducing construction costs. When glass is used as the curtain wall, a great advantage is that natural light can penetrate deeper within the building.

幕墙系统是建筑物的外部维护结构，其外壁是非结构（承重）部件，仅仅是使居住者在建筑物内免受外界天气影响。由于幕墙是非结构的，它可以由轻质材料制成，从而降低了建筑成本。当玻璃被用作幕墙时，最大的优点是自然光线能穿透进入建筑物内部深处。

[3] Cladding, as shown in Figure 8-2, is the application of one material over another to provide a skin or layer. In construction, cladding is used to provide a degree of thermal insulation and weather resistance, and to improve the appearance of buildings. Cladding can be made of any of a wide range of materials including wood, metal, brick, vinyl, and composite materials that can include aluminium, wood, blends of cement and recycled polystyrene, wheat/rice straw fibres.

覆层（图8-2）是将一种材料应用于另一种材料之上作为肤层或外层。在施工中，覆层用于提供一定程度的隔热和耐候性，并改善建筑物的外观。覆层可以由包括木材、金属、砖、乙烯基和复合材料在内的各种材料制成，复合材料可以包括铝、木材、水泥和回收聚苯乙烯的混合物、小麦/稻草纤维。

[4] The most popular (and often most expensive) glass system for application in structural glass facades is the bolted version. The glass panel requires perforations to accommodate specialized bolting hardware. Specially designed off-the-shelf hardware systems are readily available, or custom components can be designed.

用于结构玻璃外墙的最受欢迎的（通常是最昂贵的）玻璃系统是螺栓式的。玻璃面板需要打孔以适应专门的螺栓硬件。专门设计的现成硬件系统可以随时获得，也可以设计定制组件。

［5］Point-fixed clamped systems, as shown in Figure 8-11, are a solution for point fixing without the perforations in glass. In the case of a spider type fitting, the spider is rotated 45 degrees from the bolted position so that its arms align with glass seams. A thin blade penetrates through the seam between adjacent pieces of glass. An exterior plate attaches to the blade and clamps the glass in place. The bolted systems present an uninterrupted glass surface, while the clamped systems expose the small exterior clamp plate.

点固定夹挂系统（图8-11）是一种无需玻璃穿孔的点固定方法。在使用驳接件时，驳接爪应从螺栓位置旋转45度，以使其臂与玻璃接缝对齐。相邻玻璃块之间的缝隙贯穿有一薄片。驳接件外板和薄片相连并将玻璃夹挂在适当位置。点接式夹挂系统呈现不间断的玻璃表面，仅露出其小部分驳接件外板。

References 课文来源

［1］https://en.wikipedia.org/wiki/Curtain_wall_(architecture)
［2］http://www.torstencalvi.com/pages/services_1_design.shtml

Unit 9　Building Facilities

- 学习任务　建筑设施
- 教学时间　2 学时
- 学习目标　了解建筑相关设施，如给排水系统、空调系统和消防系统等。

Text 课文

Buildings contain extensive technical infrastructures, which are continually growing in complexity. The term building technology or the (DIN) standard designation technical equipment in buildings refers to all permanently installed technical equipment, inside and outside the building, designed to ensure the proper running and general use of these buildings.[1]

The term technical equipment in buildings was introduced to avoid confusion with the term building services systems used in industrial processing. Essentially, this building technology covers the following plants and installations:

—Heating, ventilation and air conditioning plants;

—Heat recovery plants;

—Piped services (plumbing);

—Energy supply and distribution;

—General building lighting;

—Window blinds systems;

—Pressure systems;

—Human conveyor systems (elevators, escalators);

—Automatic doors and gates;

—Safety and security systems (fire, intrusion protection);

—Disposal plants for sewage, flue gas, waste materials etc.

9.1 Air Conditioning Systems

The purpose of heating technology is to provide a constant and comfortable room temperature throughout the heating period. Heating technology generates hot water for space heating, and, in most plants, also for domestic use. The heating technology in a building covers: heat generation, heat distribution, and heat emission.

The generation of heat is a highly complex aspect of heating technology. In addition to conventional oil, gas, wood, or coal-fired boilers, heat is also generated using heat pumps, cogeneration plants, solar energy, or combinations of these heat producers (bivalent heat generation), and district heating transfer stations. Plumbing services are closely related to heating technology.

This covers air renewal, especially in areas such as factories, cinemas, theaters, and restaurants—in other words, in buildings where the air is used up or polluted very quickly. Despite the introduction of fresh air in this process, the room temperature must be maintained at the required level. Heating coils are used for this purpose. The majority of these are heated with hot water, although electricity or steam is sometimes used.

Our sense of well-being and efficiency is affected not only by the room temperature, but also by the humidity, cleanliness, and freshness of the air—in other words by indoor conditions tuned as finely as possible to the human organism and senses. An air conditioning plant can influence these factors. The air is treated by use of heating coils, cooling coils, and air humidifiers. Air conditioning technology today ranges from air conditioning plants for individual rooms and residential buildings through to the major plants seen for example in office buildings, shopping malls, and airports, etc.

When a new home is built, there are many different types of heating systems to consider. Some types of heating systems are forced air, radiant heat, hydronic, steam radiant, and geothermal. Each type of heat should be considered for its effectiveness in meeting the budget and heating and cooling needs for the home.

The forced air heating system is most commonly seen in residential structures. It works by heating air in a furnace and then forcing the air out into various areas of the home through installed ductwork and vents. It is also commonly known as a central heating system because it comes from a central point in the home, where it can be filtered, humidified, or dehumidified. The air can be heated with various methods, including electricity, natural gas, propane, or oil. Since it can be used to address both heating and cooling, the system is convenient for many people. [2]

The ductwork required to use this system takes space in walls, so it may be difficult to install this system in an older home, and can require extra planning with new construction. The furnace system used may be noisy and heard throughout the home. This system can also move allergens throughout the house as the air circulates. The air filtration systems will require regular maintenance to retain optimal function. This system can be expensive due to maintenance costs.

The radiant heat heating system is often praised for its ability to produce natural and comfortable heat in a home. In this system, heat is commonly delivered through a system of hot water tubes underneath the floor, although these tubes can also be installed in ceiling panels. The hot water is heated using a boiler which is usually powered by oil, natural gas, propane, or electricity. A heating stove may also be used to heat the water, powered by coal or wood. [3]

Radiant heat is often slow to heat a room because the water must first be heated and circulated through the pipes. It can be expensive to install and maintain because of the difficulty involved in getting to the tubing systems if a problem occurs with the system. Air conditioning is not available with this method, as it requires a completely separate system of ductwork.

Hydronic heat is also known as a hot water baseboard system. Much like radiant heat systems, a boiler heats hot water, which then is circulated through tubes; for hydronic heat, these tubes are located in baseboard heating units attached to the walls in each room of the home. These systems are usually quiet, energy efficient, and may be fueled by electricity, oil, or natural gas. Temperature can usually be controlled separately in each room. Baseboard units should not be blocked by curtains or furniture, making them inconvenient for some users, and as with radiant heat, hydronic systems can be slow to warm a room and require a separate cooling system. [4]

Steam radiant heating systems heat a room through upright units referred to as "radiators". These systems use either one or two pipes, and heat water through a variety of methods such as electricity, oil, or natural gas. While these units may be energy efficient and warm a room quickly, they can be inconvenient for furniture placement, as the walls and surrounding area must be clear to avoid fire hazards. Many people do not like the way radiators look in a room, and therefore choose another heating system. [5]

Geothermal heating systems are a more recent option for heating and cooling a home. These systems can be expensive to install; however, because of their ability to use the heat from the Earth to regulate temperature, they are said to greatly reduce the costs associated with heating and cooling a home. This system works for both heating and cooling because it uses the relatively constant temperature of the ground.

When a homeowner is choosing a heating system for his home, he should consider how the system will be powered in addition to how much it will cost. Considering that many of these options require separate cooling systems, it may be best to use a central heating system to combine heating with cooling in those regions where both are required. Focusing on specific needs will assist homeowners with making a decision about which system to use.

9.2 Plumbing Systems

In a house, there are four plumbing systems, as shown in Figure 9-1, enabling water to circulate: hot and cold water distribution, wastewater evacuation and pipe ventilation.

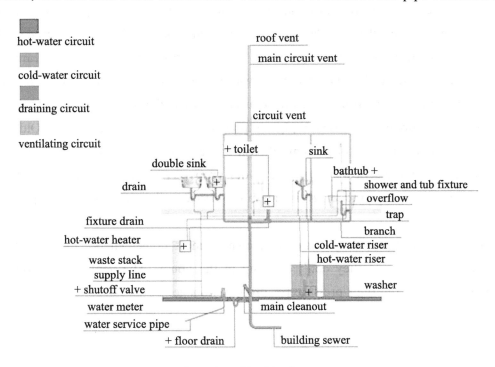

Figure 9-1 Plumbing Systems

A home's plumbing system is a complex network of water supply pipes, drainpipes, vent pipes, and more. Several different systems make up a house's plumbing. Fresh water is delivered to a home through water supply pipes from the utility or a well and is then distributed to sinks, toilets, washers, bathtubs, and related fixtures. The waste and vent system carries away used water and wastes to sewers or septic tanks.

A home's water supply system routes municipal water from the street to your house, where it branches out to deliver the water to faucets, showers, toilets, bathtubs, and appliances such as the water heater, dishwasher, and washing machine. The equipment for this delivery and distribution is essentially a system of water pipes, fittings, service valves, and faucets. These pipes and other fittings are commonly made of plastic, copper, or galvanized iron. The pipes range in diameter from 1/2 inch to 4 inches or more.

The waste system, as illustrated in Figure 9-1, is another set of pipes which carries used water away to drains and sewers. Three types of plastic can be used for external drainage and waste pipework. Acrylonitrile butadiene systems (ABS) is a very tough plastic that can be used for both hot and cold waste. It can be connected using either solvent or compression joints. Polypropylene (PP) is a softer, more flexible plastic. It is impossible to glue PP, so the connections are always made using push fit joints. The most commonly used material for external waste pipes is unplasticized polyvinyl chloride (UPVC). This type of plastic is damage resistant to most kinds of household products like bleach and washing powder. Although plastic pipes have been around for many years, until recently the most successful type has been waste pipe made from hard plastics. Plastic supply pipe (cold water) has now been introduced for use underground. Coloured blue, this medium-density polythene (MDPE) pipe is pressure and corrosion-resistant. Old mains pipes were quite often made of galvanized steel or lead, which eventually deteriorated. If you have a leaking old-style mains pipe on your property, make sure that you replace it with medium-density polythene. It is far more efficient than the old metal pipes and doesn's rot.[6]

The vent system is usually less well known to most homeowners, and its job is to ventilate sewage gases so they don't build up in the house. The vent system also helps drainpipes maintain the right pressure for proper drainage.

9.3 Fire Protection & Fire Alarm Systems

Protection from fire is critical to every home and in the event of a fire it's essential to have the right equipment so you can detect it early and protect your business assets—as well as save lives. In general, a fire alarm system is classified as either automatically actuated, manually actuated, or both. An automatic fire alarm system as illustrated in Figure 9-2 is designed to detect the unwanted presence of fire by monitoring environmental changes associated with combustion. Automatic fire alarm systems are intended to notify the building occupants to evacuate in the event of a fire or other emergency, report the event to an off-premises location in order to summon emergency services, and to prepare the structure and associated systems to control the spread of fire and smoke.[7]

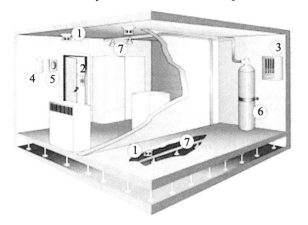

1—Smoke Detectors
2—Pull & Abort Station
3—Control Panel
4—Bell
5—Horn / Strobe Light
6—Suppression Cylinder
7—Discharge Nozzles

Figure 9-2   Fire Suppression Systems Layout

An automatic fire alarm system consists of trigger devices, fire alarm devices and other devices with auxiliary functions. It can detect physical quantities of smoke, heat, light radiation at early stages of the fire and change them into electrical signals through detectors of temperature, smoke, light and then transmit these signals into the fire alarm controller showing the site and time record of the fire.[8]

Fixed fire extinguishing/suppression systems are commonly used to protect areas containing valuable or critical equipment such as data processing rooms, telecommunication switches, and process control rooms. Their main function is to quickly extinguish a developing fire and alert occupants before extensive damage occurs by filling the protected area with a gas or chemical extinguishing agent.[9]

9.4   Intelligent Building Technology

Building technology may have to meet various requirements depending on the purpose of the building. However, the following three main requirements are prevalent:

—The human need for comfort and well-being within the building shell, tailored to specific types of building use, must be adequately met irrespective of external influences.

—The building shell must provide protection commensurate with the potential risks to protect occupants, users, and their property against damage by fire or water, damage to equipment, or attack by third parties.

—It should be possible to meet these requirements with acceptable investment costs and minimal follow-on costs for energy, operation, maintenance, and loan servicing.[10]

The relevant building technology plants can satisfy the overall requirements. We can refer to intelligent building technology (shown in Figure 9-3) when all technical equipment works optimally with regard to specific requirements.

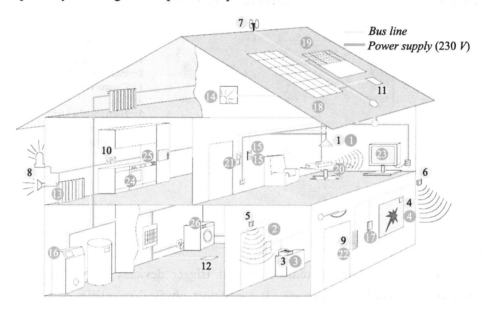

Figure 9-3　The Intelligent Building

1—Lighting control (automatic and time-based)　2—Central and group switching　3—Remote interrogation, remote control　4—Window switches　5—Occupancy detector　6—Outside surveillance system　7—Wind Speed (e. g. protection of blinds)　8—Outside siren with flashlight　9—Door locking contact　10—Socket (can be deactivated) plant　11—Rain detector (automatic closing of attic window)　12—Water detector　13—Heating valve actuators　14—Sun position-dependent blind control　15—Room temperature control　16—Condensing boiler　17—Outside sensor　18—Solar plant / photovoltaics　19—Shutter and blind control　20—Remote infrared operation　21—Operation　22—Intercom with video camera　23—TV set for monitoring and operating　24—Oven　25—Dish washer　26—Laundry machine

Glossary 词汇表

window blinds system 百叶窗系统
human conveyor system 人员输送系统
coil [kɔɪl] n. 盘卷之物、线圈
ductwork [ˈdʌktˌwɜːk] n. 管道系统
filtered [ˈfɪltəd] adj. 过滤的
humidified [hjuː(ː)ˈmɪdɪfaɪd] adj. 加湿的

dehumidified adj. 除湿的
furnace [ˈfɜːnɪs] n. 炉
baseboard [ˈbeɪsbɔːd] n. 踢脚线
venting system 通风系统，排气系统
radiant [ˈreɪdɪənt] adj. 辐射的
hydronic [haɪˈdrɒnɪk] adj. 液体循环加热（或冷却）的

geothermal [ˌdʒi(ː)əʊˈθɜːməl] *adj.*
　地热的，地温的，地热（或地温）产生的
humid [ˈhjuːmɪd] *adj.* 潮湿的，
　湿气重的，湿润的，温湿的
plumbing [ˈplʌmɪŋ] *n.* 水管装置，
　水暖工的工作，管道工程
stop valve 截止阀
cylinder [ˈsɪlɪndə] *n.* 圆筒，圆柱，汽缸，
　（尤指用作容器的）圆筒状物
contaminate [kənˈtæmɪneɪt] *v.*
　沾染，弄污，污染，使受放射性物质影
　响而无法使用
duct [dʌkt] *n.* 管道，导管，输送管，槽，
　沟，渠道

premise [ˈpremɪs] *n.* 前提，[复数] 房屋
summon [ˈsʌmən] *vt.* 传唤，召唤，
　鼓起（勇气）
accomplish [əˈkʌmplɪʃ] *vt.* 完成，达到
　（目的）
extinguishing [ɪksˈtɪŋgwɪʃɪŋ] *n.* 熄灭
suppression [səˈpreʃn] *n.* 压制，镇压，
　禁止，抑制
telecommunication [ˌtelɪkəˌmjuːnɪˈkeɪʃn]
　n. 远程通信
switch [swɪtʃ] *n.* 开关，交换机，转换器
occupant [ˈɒkjəpənt] *n.*（土地、房屋、地
　位等的）占有人，居住者

Notes 注释

[1] Buildings contain extensive technical infrastructures, which are continually growing in complexity. The term building technology or the (DIN) standard designation technical equipment in buildings refers to all permanently installed technical equipment, inside and outside the building, designed to ensure the proper running and general use of these buildings.

建筑物包含着广泛的技术基础设施，其复杂性在不断增加。建筑物中的标准指定技术设备（DIN）是指建筑物内外的所有永久性安装的技术设备，以确保这些建筑物的正常运行和使用。

[2] The forced air heating system is most commonly seen in residential structures. It works by heating air in a furnace and then forcing the air out into various areas of the home through installed ductwork and vents. It is also commonly known as a central heating system because it comes from a central point in the home, where it can be filtered, humidified, or dehumidified. The air can be heated with various methods, including electricity, natural gas, propane, or oil. Since it can be used to address both heating and cooling, the system is convenient for many people.

强制空气流动加热系统是住宅结构中使用最广泛的供暖系统，它通过在炉内加热空气并强制使得空气通过已安装的管道和通风系统进入到室内的各个区域。由于暖气都来

源于室内某集中点并在该处对空气进行过滤、加湿、去湿,因此也称为中央供暖系统。空气加热方式有许多种,包括电力、天然气、丙烷或石油。由于该系统能被用作供暖或制冷,因此它的使用对许多人来说都很方便。

[3] The radiant heat heating system is often praised for its ability to produce natural and comfortable heat in a home. In this system, heat is commonly delivered through a system of hot water tubes underneath the floor, although these tubes can also be installed in ceiling panels. The hot water is heated using a boiler which is usually powered by oil, natural gas, propane, or electricity. A heating stove may also be used to heat the water, powered by coal or wood.

辐射热供暖系统经常因其能够在家中产生自然而又舒适的热量而广受称赞。在该系统中,热量通过位于地板下的热水管系统来传递,这些管道也可以安装在天花板中。热水一般由石油、天然气、丙烷或电力的锅炉来加热,有时也可由使用煤和木头的火炉来加热。

[4] Hydronic heat is also known as a hot water baseboard system. Much like radiant heat systems, a boiler heats hot water, which then is circulated through tubes; for hydronic heat, these tubes are located in baseboard heating units attached to the walls in each room of the home. These systems are usually quiet, energy efficient, and may be fueled by electricity, oil, or natural gas. Temperature can usually be controlled separately in each room. Baseboard units should not be blocked by curtains or furniture, making them inconvenient for some users, and as with radiant heat, hydronic systems can be slow to warm a room and require a separate cooling system.

液体循环加热也被称为基于护壁板的热水系统。和辐射热系统相类似,加热器加热热水后在管道内循环。在该系统中,管道位于连接到家中每个房间墙上的护壁板加热单元中。该系统通常比较安静、高效节能并可以由电力、石油或天然气来供应能量。每个房间的温度都可以独立控制。由于护壁板部分不能被窗帘或家具所遮挡,这给一些使用者带来了不便,而且和辐射热系统类似,液体循环加热式供暖速度慢并需要一个独立制冷系统配合。

[5] Steam radiant heating systems heat a room through upright units referred to as "radiators". These systems use either one or two pipes, and heat water through a variety of methods such as electricity, oil, or natural gas. While these units may be energy efficient and warm a room quickly, they can be inconvenient for furniture placement, as the walls and surrounding area must be clear to avoid fire hazards. Many people do not like the way radiators look in a room, and therefore choose another heating system.

蒸气辐射供暖系统通过一个直立的散热部件给房间供暖。该系统可以使用一根或两根管道，加热方式多样，可使用电力、石油或天然气。尽管蒸气辐射供暖系统可以节能且升温速度快，但室内家具布置不方便，墙体附近不能有易燃物，以避免火灾危险，并需要一个独立制冷系统配合。大多数人不太喜欢室内直立的散热部件，转而选择其他的供暖方式。

[6] The waste system as illustrated in Figure 9-1, is another set of pipes which carries used water away to drains and sewers. Three types of plastic can be used for external drainage and waste pipework. Acrylonitrile butadiene systems (ABS) is a very tough plastic that can be used for both hot and cold waste. It can be connected using either solvent or compression joints. Polypropylene (PP) is a softer, more flexible plastic. It is impossible to glue PP, so the connections are always made using push fit joints. The most commonly used material for external waste pipes is unplasticized polyvinyl chloride (UPVC). This type of plastic is damage resistant to most kinds of household products like bleach and washing powder. Although plastic pipes have been around for many years, until recently the most successful type has been waste pipe made from hard plastics. Plastic supply pipe (cold water) has now been introduced for use underground. Coloured blue, this medium-density polythene (MDPE) pipe is pressure and corrosion-resistant. Old mains pipes were quite often made of galvanized steel or lead, which eventually deteriorated. If you have a leaking old-style mains pipe on your property, make sure that you replace it with medium-density polythene. It is far more efficient than the old metal pipes and doesn't rot.

废水系统（图9-1）是另一类管道系统，主要把用过的废水带到下水道和污水管。目前有三种类型的塑料可以用作排水管和废水管系统。丁腈橡胶（ABS）是一种非常结实的塑料，能被用作热或冷废水管道，它可以用溶剂或压力接合进行连接。聚丙烯（PP）是一种比较软、更具柔韧性的塑料，PP不可能粘合，所以连接通常都采用推入式接头。外置废水管的最常用材料是未增塑聚氯乙烯（UPVC），这种材料对大多数家用漂白剂和洗涤粉剂具有抗破坏力。尽管塑料管道已经存在了许多年，但直到最近，最为成功的仍然是由硬塑料制成的废水管道。塑料供水管（冷水）现在已经引入地下使用。天蓝色中密度聚乙烯管（MDPE）耐压力耐腐蚀。旧的主管道通常由镀锌钢管或铅管制成，最终会腐坏，如果发生泄漏，可以用中密度聚乙烯管来代替，这种材料更有效率且耐腐蚀。

[7] Protection from fire is critical to every home and in the event of a fire it's essential to have the right equipment so you can detect it early and protect your business assets—as well as save lives. In general, a fire alarm system is classified as either automatically actuated, manually actuated, or both. An automatic fire alarm system as illustrated in Figure 9-2 is designed to detect the unwanted presence of fire by monitoring

environmental changes associated with combustion. Automatic fire alarm systems are intended to notify the building occupants to evacuate in the event of a fire or other emergency, report the event to an off-premises location in order to summon emergency services, and to prepare the structure and associated systems to control the spread of fire and smoke.

防火对每个家庭都很重要,在着火时,需要有合适的设备以便及早发现火灾并保护商业财产以及挽救生命。总的来讲,火灾报警系统一般可以分为自动报警、人工报警或两者兼备。图9-2所示的自动火灾报警系统的设计是通过监测和燃烧相关的环境变化来探测意外之火。其目的是能通知业主在发生火灾或紧急情况时疏散,把事件报告给外部以召集紧急服务,并准备结构以及相关的系统来控制火灾和烟雾的扩散。

[8] An automatic fire alarm system consists of trigger devices, fire alarm devices and other devices with auxiliary functions. It can detect physical quantities of smoke, heat, light radiation at early stages of the fire and change them into electrical signals through detectors of temperature, smoke, light and then transmit these signals into the fire alarm controller showing the site and time record of the fire.

火灾自动报警系统由触发装置、火灾警报装置以及其他具有辅助功能的装置组成。它能够在火灾初期发现大量的烟雾、热量和光辐射,并通过感温、感烟和感光探测器将它们转换成电信号传输到火灾报警控制器,并同时显示出火灾发生的部位,记录火灾发生的时间。

[9] Fixed fire extinguishing/suppression systems are commonly used to protect areas containing valuable or critical equipment such as data processing rooms, telecommunication switches, and process control rooms. Their main function is to quickly extinguish a developing fire and alert occupants before extensive damage occurs by filling the protected area with a gas or chemical extinguishing agent.

固定式灭火系统通常用来保护有价值的或至关重要的设备区域,如数据处理室、远程通信交换机和程序控制室。它们主要的功能是通过在保护区内注入气体或化学灭火剂从而迅速扑灭起火点并提醒使用者以防止更大的损失发生。

[10] Building technology may have to meet various requirements depending on the purpose of the building. However, the following three main requirements are prevalent:

——The human need for comfort and well-being within the building shell, tailored to specific types of building use, must be adequately met irrespective of external influences.

——The building shell must provide protection commensurate with the potential risks to protect occupants, users, and their property against damage by fire or water, damage to equipment, or attack by third parties.

—It should be possible to meet these requirements with acceptable investment costs and minimal follow-on costs for energy, operation, maintenance, and loan servicing.

建筑技术可能需要根据建筑物的目的来满足各种要求。然而，以下三个主要要求是普遍存在的：

——无论外部影响如何，必须充分满足人们对建筑物内的舒适和健康的需求，以适合特定类型的建筑物使用。

——建筑物必须提供与潜在风险相适应的保护措施，以保护住户、使用者及他们的财产免遭火灾或水灾的破坏、设备的损坏或第三方的袭击。

——应该能够以可接受的投资成本和能源、运营、维修和贷款服务的最低后续成本来满足这些要求。

References 课文来源

[1] http://www.linkedin.com/pulse/bms-benefits-control-systems-strategies-building-intelligent-plus? trk=mp-reader-card

[2] http://www.visualdictionaryonline.com/house/plumbing/plumbing-system.php

[3] https://www.hometips.com/plumbing_fixtures.html

[4] http://kerberossecurity.com/? page_id=427

[5] http://www.linkedin.com/pulse/bms-benefits-control-systems-strategies-building-intelligent-plus? trk=mp-reader-card

Unit 10　Construction Process and Technology

> ▶ 学习任务　施工过程和技术
> ▶ 教学时间　2 学时
> ▶ 学习目标　初步了解施工工艺的类型和技术特点

Text 课文

Construction is the process of constructing a building or infrastructure. Construction differs from manufacturing in that manufacturing typically involves mass production of similar items without a designated purchaser, while construction typically takes place on location for a known client. Construction as an industry comprises six to nine percent of the gross domestic product of developed countries. Construction starts with planning, design, and financing; it continues until the project is built and ready for use.[1]

Large-scale construction requires collaboration across multiple disciplines. An architect normally manages the job, and a construction manager, design engineer, construction engineer or project manager supervises it. For the successful execution of a project, effective planning is essential. Those involved with the design and execution of the infrastructure in question must consider zoning requirements, the environmental impact of the job, the successful scheduling, budgeting, construction-site safety, availability and transportation of building materials, logistics, inconvenience to the public caused by construction delays and bidding, etc. The largest construction projects are referred to as megaprojects.

10.1 Types of Construction

In general, there are three sectors of construction: buildings, infrastructure and industrial. Building construction is usually further divided into residential and non-residential (commercial/institutional). Infrastructure is often called heavy/highway, heavy civil or heavy engineering. It includes large public works, dams, bridges, highways, water/wastewater and utility distribution. Industrial includes refineries, process chemical, power generation, mills and manufacturing plants. There are other ways to break the industry into sectors or markets.

Building construction is the process of adding structure to real property or construction of buildings. The majority of building construction jobs are small renovations, such as addition of a room, or renovation of a bathroom. Often, the owner of the property acts as laborer, paymaster, and design team for the entire project. Although building construction projects typically include various common elements, such as design, financial, estimating and legal considerations, many projects of varying sizes reach undesirable end results, such as structural collapse, cost overruns, and/or litigation. For this reason, those with experience in the field make detailed plans and maintain careful oversight during the project to ensure a positive outcome.

Commercial building construction is procured privately or publicly utilizing various delivery methodologies, including cost estimating, hard bid, negotiated price, traditional, management contracting, construction management-at-risk, design & build and design-build bridging.

Residential construction practices, technologies, and resources must conform to local building authority regulations and codes of practice. Materials readily available in the area generally dictate the construction materials used (e. g. brick versus stone, versus timber). Cost of construction on a per square meter (or per square foot) basis for houses can vary dramatically based on site conditions, local regulations, economies of scale (custom designed homes are often more expensive to build) and the availability of skilled tradespeople. As residential construction (as well as all other types of construction) can generate a lot of waste, careful planning again is needed here.

The most popular method of residential construction in North America is wood-framed construction. Typical construction steps for a single-family or small multi-family house are:

—Obtain an engineered soil test of lot where construction is planned;

—Develop floor plans and obtain a materials list for estimations (more recently performed with estimating software);

—Obtain structural engineered plans for foundation;

—Obtain lot survey;

—Obtain government building approval if necessary;

—Clear the building site;

—Survey to stake out for the foundation;

—Excavate the foundation and dig footers;

—Install plumbing grounds;

—Pour a foundation and footers with concrete;

—Build the main load - bearing structure out of thick pieces of wood and possibly metal I - beams for large spans with few supports;

—Add floor and ceiling joists and install subfloor panels;

—Cover outer walls and roof in plywood and a water-resistive barrier;

—Install roof shingles or other covering for flat roof;

—Cover the walls with siding, typically vinyl, wood, or brick veneer but possibly stone or other materials;

—Install windows;

—Add internal plumbing, HVAC, electrical, and natural gas utilities;

—Install bathroom fixtures;

—Spackle, prime, and paint interior walls and ceilings;

—Additional tiling on top of cementboard for wet areas, such as the bathroom and kitchen backsplash;

—Installation of final floor covering, such as floor tile, carpet, or wood flooring;

—Installation of major appliances.[2]

10.2 Additive Building Construction

New techniques of building construction are being researched, made possible by advances in 3D printing technology. In a form of additive building construction, similar to the additive manufacturing techniques for manufactured parts (shown in Figure 10-1), building printing is making it possible to flexibly construct small commercial buildings and private habitations in around 20 hours, with built-in plumbing and electrical facilities, in one continuous build (shown in Figure 10-2) using large 3D printers.[3]

Figure 10-1 3D Printer Builds Truss

Figure 10-2 Giant Chinese 3D Printer Builds 10 Houses in just One Day

Current machines are being integrated into automated and semi-automated production lines and, because of the scale of construction, will feature elements of additive, subtractive and formative manufacturing processes, to handle material deposition at one scale and finishing at another. Because of the cost, 3D printing at construction scales demands clever design and can respond to the demands of architects and engineers for high value, high performance building components. Potential advantages of these technologies include faster construction, lower labor costs, increased complexity and/or accuracy, greater integration of function and less waste produced. There are a variety of 3D printing methods used at construction scale; these include the following main methods: extrusion (concrete/cement, wax, foam, polymers), powder bonding (polymer bond, reactive bond, sintering) and additive welding. 3D printing at a construction scale will have a wide variety of applications within the private, commercial, industrial and public sectors. Development has been slow and sporadic, since its development in the mid 1990s, where initially it was explored as a scaled version of mainstream 3D printing, having both novelty value and early research funding in both the US and Europe. The term 'Construction 3D Printing' was first coined by James B Gardiner in 2011. Working versions of 3D-printing building technology are already printing 2 metres (6 ft 7 in) of building material per hour as of January 2013, with the next-generation printers capable of 3.5 metres (11 ft) per hour, sufficient to complete a building in a week. The first residential building in Europe, constructed using the 3D printing construction technology, was the home in Yaroslavl (Russia) with the area of 2985 sq. m. The walls of the building were printed by the company SPECAVIA in December 2015.

A number of different approaches have been demonstrated to date which include on-site and off-site fabrication of buildings and construction components, using industrial robots, gantry systems and tethered autonomous vehicles. Demonstrations of construction 3D printing technologies to date have included fabrication of housing, construction components (cladding and structural panels and columns), bridges, artificial reefs, follies and sculptures. Current efforts focus on integrating the advantages of digital fabrication within factory based construction manufacturing. Stand alone and on-site machines are in planning and research, ranging from modified autonomous concrete/gypsum/mineral paste pumping/spraying, composite fiber spinning and ultimately swarm construction agents, where construction 3D printing merges with robotics and AI systems. Pilot studies have demonstrated that construction 3D printing may be well suited for construction of extraterrestrial structures on the Moon or other planets, where environmental conditions are less conducive to human labor-intensive building practices.

10.3 Sustainable Construction

In the current trend of sustainable construction, the recent movements of New Urbanism and New Classical Architecture promote a sustainable approach towards construction, that appreciates and develops smart growth, architectural tradition and classical design. This is in contrast to modernist and short-lived globally uniform architecture, as well as opposing solitary housing estates and suburban sprawl. Both trends started in the 1980s.

Every engineering discipline is engaged in sustainable design, employing numerous initiatives, especially life cycle analysis (LCA), pollution prevention, design for the environment (DfE), design for disassembly (DfD), and design for recycling (DfR). These are replacing or at least changing pollution control paradigms. For example, concept of a "cap and trade" has been tested and works well for some pollutants. This is a system where companies are allowed to place a "bubble" over a whole manufacturing complex or trade pollution credits with other companies in their industry instead of a "stack-by-stack" and "pipe-by-pipe" approach, i.e. the so-called "command and control" approach. Such policy and regulatory innovations call for some improved technology based approaches as well as better quality-based approaches, such as leveling out the pollutant loadings and using less expensive technologies to remove the first large bulk of pollutants, followed by higher operation and maintenance (O&M) technologies for the more difficult to treat stacks and pipes. But, the net effect can be a greater reduction of pollutant emissions and effluents than treating each stack or pipe as an independent entity. This is a foundation for most sustainable design approaches, i.e. conducting a life-cycle analysis, prioritizing the most important problems, and matching the technologies and operations to address them. The problems will vary by size (e.g. pollutant loading), difficulty in treating, and feasibility. The most intractable problems are often those that are small but very expensive and difficult to treat, i.e. less feasible. Of course, as with all paradigm shifts, expectations must be managed from both a technical and an operational perspective. Historically, sustainability considerations have been approached by engineers as constraints on their designs. For example, hazardous substances generated by a manufacturing process were dealt with as a waste stream that must be contained and treated. The hazardous waste production had to be constrained by selecting certain manufacturing types, increasing waste handling facilities, and if these did not entirely do the job, limiting rates of production. Green engineering recognizes that these processes are often inefficient economically and environmentally, calling for a comprehensive, systematic life cycle approach. Green engineering attempts to achieve four goals:

—Waste reduction;

—Materials management;

—Pollution prevention;

—Product enhancement.

Green engineering, as shown in Figure 10-3, encompasses numerous ways to improve processes and products to make them more efficient from an environmental and sustainable standpoint. Every one of these approaches depends on viewing possible impacts in space and time. [4] The design must consider short and long-term impacts. Those impacts beyond the near-term are the province of sustainable design. The effects may not manifest themselves for decades. In the mid-twentieth century, designers specified the use of what are now known to be hazardous building materials, such as asbestos flooring, pipe wrap and shingles, lead paint and pipes, and even structural and mechanical systems that may have increased the exposure to molds and radon.[5] Those decisions have led to risks to people inhabiting these buildings. It is easy in retrospect to criticize these decisions, but many were made for noble reasons, such as fire prevention and durability of materials.

Figure 10-3　An Autonomous and Mobile Station
That Replenishes Energy for Electric Vehicles Using Solar Energy

However, it does illustrate that seemingly small impacts when viewed through the prism of time can be amplified exponentially in their effects. Sustainable design requires a complete assessment of a design in place and time. Some impacts may not occur until centuries in the future. For example, the extent to which we decide to use nuclear power to generate electricity is a sustainable design decision. The radioactive wastes may have half-lives of hundreds of thousands of years. That is, it will take all these years for half of the radioactive isotopes to decay. Radioactive decay is the spontaneous transformation of one element into another. This occurs by irreversibly changing the number of protons in the nucleus. Thus, sustainable designs of such enterprises must consider highly uncertain futures. For example, even if we properly place warning signs about these hazardous wastes, we do not know if the English language will be understood. All four goals of green engineering mentioned above are supported by a long-term, life cycle point of view. A life cycle analysis is a holistic approach to consider the entirety of a product, process or activity, encompassing raw materials, manufacturing, transportation, distribution, use, maintenance, recycling, and final disposal. In other words, assessing its life cycle should yield a complete picture of the product. The first step in a life cycle assessment is to gather data on the flow of a material through an identifiable society. Once the quantities of various components of such a flow are known, the important functions and impacts of each step in the production, manufacture, use, and recovery/disposal are estimated. Thus, in sustainable design, engineers must optimize for variables that give the best performance in temporal frames.

Glossary 词汇表

designated ['dezɪgneɪtɪd] *adj.* 指定的，选定的
methodology [ˌmeθə'dɒlədʒi] *n.* 方法论
stake out 放样
excavate ['ekskəveɪt] *vt.* 开挖
joist [dʒɔɪst] *n.* 龙骨
additive building construction 增材法施工
architect ['ɑːkɪtekt] *n.* 建筑师
extrusion [ɪk'struːʒn] *n.* 挤出；推出；赶出；喷出
powder bonding 粉末粘合
sporadic [spə'rædɪk] *adj.* 偶发的
sculpture ['skʌlptʃə(r)] *n.* 雕像
gypsum ['dʒɪpsəm] *n.* 石膏
scaffolded *n.* 脚手架
encompass [ɪn'kʌmpəs] *vt.* 围绕，包围
sustainable standpoint *n.* 可持续发展的观点

Notes 注释

[1] Construction is the process of constructing a building or infrastructure. Construction differs from manufacturing in that manufacturing typically involves mass production of similar items without a designated purchaser, while construction typically takes place on location for a known client. Construction as an industry comprises six to nine percent of the gross domestic product of developed countries. Construction starts with planning, design, and financing; it continues until the project is built and ready for use.

施工是建造建筑物或基础设施的过程。施工不同于制造，制造通常涉及没有指定购买者的情况下生产类似物品，而施工通常在已知客户位置的情况下进行。建筑业占发达国家国内生产总值的6%至9%。建筑施工从规划、设计和融资开始，并持续到项目建成并准备投入使用为止。

[2] Typical construction steps for a single-family or small multi-family house are:
——Obtain an engineered soil test of lot where construction is planned;
——Develop floor plans and obtain a materials list for estimations (more recently performed with estimating software);
——Obtain structural engineered plans for foundation;
——Obtain lot survey;
——Obtain government building approval if necessary;

—Clear the building site;

—Survey to stake out for the foundation;

—Excavate the foundation and dig footers;

—Install plumbing grounds;

—Pour a foundation and footers with concrete;

—Build the main load-bearing structure out of thick pieces of wood and possibly metal I-beams for large spans with few supports;

—Add floor and ceiling joists and install subfloor panels;

—Cover outer walls and roof in plywood and a water-resistive barrier;

—Install roof shingles or other covering for flat roof;

—Cover the walls with siding, typically vinyl, wood, or brick veneer but possibly stone or other materials;

—Install windows;

—Add internal plumbing, HVAC, electrical, and natural gas utilities;

—Install bathroom fixtures;

—Spackle, prime, and paint interior walls and ceilings;

—Additional tiling on top of cementboard for wet areas, such as the bathroom and kitchen backsplash;

—Installation of final floor covering, such as floor tile, carpet, or wood flooring;

—Installation of major appliances.

单体家庭或小型多户住宅的典型施工步骤是：

——提供计划施工地的工程土工试验报告。

——平面施工图出图并获得材料清单以进行估算（用估算软件最近进行）。

——提供基础的结构工程方案。

——提供地质勘探报告。

——如有必要获得政府批准。

——清除建筑工地。

——基础放样勘察。

——开挖地基、柱基。

——安装管道地下设施。

——用混凝土浇筑基础和柱基。

——用厚木板和尽可能的金属工字梁（少支撑的大跨度梁）建造主要承重结构。

——添加地板和吊顶龙骨，安装地板底板。

——在胶合板和防水屏障上覆盖外墙和屋顶。

——安装屋顶瓦或平屋顶的其他覆盖物。

——用墙板覆盖墙壁，通常是乙烯基、木材或砖饰面，但可能是石头或其他材料。

——安装窗户。

——安装内部管道、暖通空调、电气和天然气设施。

——安装浴室固定装置。

——腻子、底漆和内墙和天花板油漆。

——在潮湿区域的水泥板顶部贴瓷砖，如浴室和厨房防溅墙。

——安装地板面层，如地砖、地毯或木地板。

——安装主要家用电器。

[3] New techniques of building construction are being researched, made possible by advances in 3D printing technology. In a form of additive building construction, similar to the additive manufacturing techniques for manufactured parts (shown in Figure 10-1), building printing is making it possible to flexibly construct small commercial buildings and private habitations in around 20 hours, with built-in plumbing and electrical facilities, in one continuous build (shown in Figure 10-2) using large 3D printers.

建筑施工的新技术正在研究中，3D打印技术的进步使其成为可能。在增材成型的建筑施工工艺中，类似于制造零件的增材制造技术（图10-1），建筑打印使得在大约20小时左右灵活地建造小型商业建筑和私人住宅成为可能（图10-2），并使用大型3D打印机内置的管道和电气设施连续建造。

[4] Green engineering encompasses numerous ways to improve processes and products to make them more efficient from an environmental and sustainable standpoint. Every one of these approaches depends on viewing possible impacts in space and time.

绿色工程包含了许多方法来改进流程和产品，使其从环境和可持续的角度上来说更有效率。这些方法中的每一种都依赖于观察空间和时间可能产生的影响。建筑师应考虑场所感。工程师应将现场布置视为一组跨界的不稳定的状态。

[5] The design must consider short and long-term impacts. Those impacts beyond the near-term are the province of sustainable design. The effects may not manifest themselves for decades. In the mid-twentieth century, designers specified the use of what are now known to be hazardous building materials, such as asbestos flooring, pipe wrap and shingles, lead paint and pipes, and even structural and mechanical systems that may have increased the exposure to molds and radon.

设计必须考虑短期和长期的影响。这些超出近期的影响是可持续设计的所在。这些影响可能不会在几十年内显现出来。在二十世纪中叶，设计师们指定使用现在已知的危险建筑材料，例如石棉地板、管道包装和木瓦、含铅油漆和管道，甚至是可能增加接触模具和氡气暴露的结构和机械系统。

References 课文来源

[1] https://en.wikipedia.org/wiki/Construction
[2] https://en.wikipedia.org/wiki/Construction_3D_printing
[3] https://en.wikipedia.org/wiki/Sustainable_design
[4] https://en.wikipedia.org/wiki/Sustainable_engineering

Unit 11 Construction Management

> ▶▶ 学习任务　施工管理
> ▶▶ 教学时间　2学时
> ▶▶ 学习目标　了解施工管理的内容和项目启动、执行以及后期管理方面的技术要点。

Text 课文

Construction project management (CM) is a professional service that uses specialized, project management techniques to oversee the planning, design, and construction of a project, from its beginning to its end. The purpose of CM is to control a project's time, cost and quality. CM is compatible with all project delivery systems, including design-bid-build, design-build, CM At-Risk and Public Private Partnerships.[1] Every construction project features some amount of CM. However, professional construction managers, or CMs, are typically reserved for lengthy, large-scale, high budget undertakings (commercial real estate, transportation infrastructure, industrial facilities, military infrastructure, etc...), called capital projects. No matter the setting, a CM's responsibility is to the owner, and to a successful project.

A contractor is assigned to a construction project during the design or once the design has been completed by a licensed architect. This is done by going through a bidding process with different contractors. The contractor is selected by using one of three common selection methods: low-bid selection, best-value selection, or qualifications-based selection.

A construction manager should have the ability to handle public safety, time management, cost management, quality management, decision making, mathematics, working drawings, and human resources.

11.1 Obtaining the Project

11.1.1 Bids

A bid is given to the owner by construction managers that are willing to complete their construction project. A bid tells the owner how much money they should expect to pay the construction management company in order for them to complete the project.

—Open bid: An open bid is used for public projects. Any and all contractors are allowed to submit their bid due to public advertising.

—Closed bid: A closed bid is used for private projects. A selection of contractors is sent an invitation for bid so only they can submit a bid for the specified project.[2]

11.1.2 Selection methods

—Low-bid selection: This selection focuses on the price of a project. Multiple construction management companies submit a bid to the owner that is the lowest amount they are willing to do the job for. Then the owner usually chooses the company with the lowest bid to complete the job for them.

—Best-value selection: This selection is focuses on both the price and qualifications of the contractors submitting bids. This means that the owner chooses the contractor with the best price and the best qualifications. The owner decides by using a request for proposal (RFP), which provides the owner with the contractor's exact form of scheduling and budgeting that the contractor expects to use for the project.

—Qualifications-based selection: This selection is used when the owner decides to choose the contractor only on the basis of their qualifications. The owner then uses a request for qualifications (RFQ), which provides the owner with the contractor's experience, management plans, project organization, and budget and schedule performance. The owner may also ask for safety records and individual credentials of their members. This method is most often used when the contractor is hired early during the design process so that the contractor can provide input and cost estimates as the design develops.

11.1.3 Payment contracts

—Lump sum: This is the most common type of contract. The construction manager and the owner agree on the overall cost of the construction project and the owner is responsible for paying that amount whether the construction project exceeds or falls below the agreed price of payment.

—Cost plus fee: This contract provides payment for the contractor including the total cost of the project as well as a fixed fee or percentage of the total cost. This contract is beneficial to the contractor since any additional costs will be paid for, even though they were unexpected for the owner.

—Guaranteed maximum price: This contract is the same as the cost-plus-fee contract although there is a set price that the overall cost and fee do not go above.

—Unit price: This contract is used when the cost cannot be determined ahead of time. The owner provides materials with a specific unit price to limit spending.

11.2 Project Stages

11.2.1 Design

The design stage involves four steps: programming and feasibility, schematic design, design development, and contract documents. It is the responsibility of the design team to ensure that the design meets all building codes and regulations. It is during the design stage that the bidding process takes place.

—Conceptual/Programming and feasibility: The needs, goals, and objectives must be determined for the building. Decisions must be made on the building size, number of rooms, how the space will be used, and who will be using the space. This must all be considered to begin the actual designing of the building. This phase is normally a written list of each room or space, the critical information about those spaces, and the approximate square footage of each area.

—Schematic design: Schematic designs are sketches used to identify spaces, shapes, and patterns. Materials, sizes, colors, and textures must be considered in the sketches. This phase usually involves developing the floor plan, elevations, a site plan, and possibly a few details.

—Design development: This step requires research and investigation into what materials and equipment will be used as well as their cost. During this phase, the drawings are refined with information from structural, plumbing, mechanical, and electrical engineers. It also involves a more rigorous evaluation how the applicable building codes will impact the project.

—Contract documents: Contract documents are the final drawings and specifications of the construction project. They are used by contractors to determine their bid while builders use them for the construction process. Contract documents can also be called working drawings.

11.2.2 Pre-construction

The pre-construction stage begins when the owner gives a notice to proceed to the contractor that they have chosen through the bidding process. A notice to proceed is when the owner gives permission to the contractor to begin their work on the project. The first step is to assign the project team which includes the project manager (PM), contract administrator, superintendent, and field engineer.

—Project manager: The project manager is in charge of the project team.

—Contract administrator: The contract administrator assists the project manager as well as the superintendent with the details of the construction contract.

—Superintendent: It is the superintendent's job to make sure everything is on schedule including flow of materials, deliveries, and equipment. They are also in charge of coordinating on-site construction activities.

—Field engineer: A field engineer is considered an entry-level position and is responsible for paperwork.

During the pre-construction stage, a site investigation must take place. A site investigation takes place to discover if any steps need to be implemented on the job site. This is in order to get the site ready before the actual construction begins. This also includes any unforeseen conditions such as historical artifacts or environment problems. A soil test must be done to determine if the soil is in good condition to be built upon.

11.2.3 Procurement

The procurement stage is when labor, materials and equipment needed to complete the project are purchased. This can be done by the general contractor if the company does all their own construction work. If the contractor does not do their own work, they obtain it through subcontractors. Subcontractors are contractors who specialize in one particular aspect of the construction work such as concrete, welding, glass, or carpentry. Subcontractors are hired the same way a general contractor would be, which is through the bidding process. Purchase orders are also part of the procurement stage.[3]

Purchase orders: A purchase order is used in various types of businesses. In this case, a purchase order is an agreement between a buyer and seller that the products purchased meet the required specifications for the agreed price.

11.2.4 Construction

The construction stage begins with a pre-construction meeting brought together by the superintendent. The pre-construction meeting is meant to make decisions dealing with work hours, material storage, quality control, and site access. The next step is to move everything onto the construction site and set it all up.

A Contractor progress payment schedule is a schedule of when (according to project milestones or specified dates) contractors and suppliers will be paid for the current progress of installed work.

Progress payments are partial payments for work completed during a portion, usually a month, during a construction period. Progress payments are made to general contractors, subcontractors, and suppliers as construction projects progress. Payments are typically made on a monthly basis but could be modified to meet certain milestones. Progress payments are an important part of contract administration for the contractor. Proper preparation of the information necessary for payment processing can help the contractor financially complete the project.

11.2.5 Owner occupancy

Once the owner moves into the building, a warranty period begins. This is to ensure that all materials, equipment, and quality meet the expectations of the owner that are included within the contract.

11.3 Issues Resulting from Construction

11.3.1 Dust and mud

When construction vehicles are driving around a site or moving earth, a lot of dust is created, especially during the dryer months. This may cause disruption for surrounding businesses or homes. A popular method of dust control is to have a water truck driving through the site spraying water on the dry dirt to minimize the movement of dust within and out of the construction site. When water is introduced mud is created. This mud sticks to the tires of the construction vehicles and is often lead out to the surrounding roads. A good practice is to have a street sweeper clean the roads at least once a day to minimize dirty road conditions. [4]

11.3.2 Environmental protections

—Storm water pollution: As a result of construction, the soil is displaced from its original location which can possibly cause environmental problems in the future. Runoff can occur during storms which can possibly transfer harmful pollutants through the soil to rivers, lakes, wetlands, and coastal waters.

—Endangered species: If endangered species have been found on the construction site, the site must be shut down for some time. The construction site must be shut down for as long as it takes for authorities to make a decision on the situation. Once the situation has been assessed, the contractor makes the appropriate accommodations to not disturb the species.

—Vegetation: There may often be particular trees or other vegetation that must be protected on the job site. This may require fences or security tape to warn builders that they must not be harmed.

—Wetlands: The contractor must make accommodations so that erosion and water flow are not affected by construction. Any liquid spills must be maintained due to contaminants that may enter the wetland.

—Historical or cultural artifacts: Artifacts may include arrowheads, pottery shards, and bones. All work comes to a halt if any artifacts are found and will not resume until they can be properly examined and removed from the area.

11.4 Construction Activity Documentation

Project meetings take place at scheduled intervals to discuss the progress on the construction site and any concerns or issues. The discussion and any decisions made at the meeting must be documented. Diaries, logs, and daily field reports keep track of the daily activities on a job site each day.

—Diaries: Each member of the project team is expected to keep a project diary. The diary contains summaries of the day's events in the member's own words. They are used to keep track of any daily work activity, conversations, observations, or any other relevant information regarding the construction activities. Diaries can be referred to when disputes arise and a diary happens to contain information connected with the disagreement. Diaries that are handwritten can be used as evidence in court.

—Logs: Logs keep track of the regular activities on the job site such as phone logs, transmittal logs, delivery logs, and RFI (Request for Information) logs.

—Daily field reports: Daily field reports are a more formal way of recording information on the job site. They contain information that includes the day's activities, temperature and weather conditions, delivered equipment or materials, visitors on the site, and equipment used that day.

Labor statements are required on a daily basis and list of labor are needed for labor planning to complete a project in time.

11.5 Resolving Disputes

—Mediation: Mediation uses a third party mediator to resolve any disputes. The mediator helps both disputing parties to come to a mutual agreement. This process ensures that no attorneys become involved in the dispute and is less time-consuming.

—Minitrial: A minitrial takes more time and money than a mediation. The minitrial takes place in an informal setting and involves some type of advisor or attorney that must be paid. The disputing parties may come to an agreement or the third party advisor may offer their advice. The agreement is nonbinding and can be broken.

—Arbitration: Arbitration is the most costly and time-consuming way to resolve a dispute. [5] Each party is represented by an attorney while witnesses and evidence are presented. Once all information is provided on the issue, the arbitrator makes a ruling which provides the final decision. The arbitrator provides the final decision on what must be done and it is a binding agreement between each of the disputing parties.

11.6　Business Model

The construction industry typically includes three parties: an owner, a licensed designer (architect or engineer) and a builder (usually known as a general contractor). There are traditionally two contracts between these parties as they work together to plan, design and construct the project. The first contract is the owner-designer contract, which involves planning, design, and construction contract administration. The second contract is the owner-contractor contract, which involves construction. An indirect third-party relationship exists between the designer and the contractor, due to these two contracts.[6]

An owner may also contract with a construction project management company as an adviser, creating a third contract relationship in the project. The construction manager's role is to provide construction advice to the designer, design advice to the constructor on the owner's behalf and other advice as necessary.

11.6.1　Design, bid, build contracts

The phrase "design, bid, build" describes the prevailing model of construction management, in which the general contractor is engaged through a tender process after designs have been completed by the architect or engineer.

11.6.2　Design-build contracts

Many owners—particularly government agencies—let out contracts known as design-build contracts. In this type of contract, the construction team (known as the design-builder) is responsible for taking the owner's concept and completing a detailed design before (following the owner's approval of the design) proceeding with construction. Virtual design and construction technology may be used by contractors to maintain a tight construction time.

There are three main advantages to a design-build contract. First, the construction team is motivated to work with the architect to develop a practical design. The team can find creative ways to reduce construction costs without reducing the function of the final product. The second major advantage involves the schedule. Many projects are commissioned within a tight time frame. Under a traditional contract, construction cannot begin until after the design is finished and the project has been awarded to a bidder. In a design-build contract the contractor is established at the outset, and construction activities can proceed concurrently with the design. The third major advantage is that the design-build contractor has an incentive to keep the combined design and construction costs within the owner's budget. If speed is important, design and construction contracts can be awarded separately; bidding takes place on preliminary plans in a not-to-exceed contract instead of a single, firm design-build contract.

The major problem with design-build contracts is an inherent conflict of interest. In a standard contract the architect works for the owner and is directly responsible to the owner. In design-build the architect works for the design-builder, not the owner, therefor the design-builder may make design and construction decisions that benefit the design-builder, but that do not benefit the owner. During construction, the architect normally acts as the owner's representative. This includes reviewing the builder's work and ensuring that the products and methods meet specifications and codes. The architect's role is compromised when the architect works for the design-builder and not for the owner directly. Thus, the owner may get a building that is over-designed to increase profits for the design-builder, or a building built with lesser-quality products to maximize profits.

Glossary 词汇表

open bid 公开招标
closed bid 非公开招标
low-bid selection 低价中标
best-value selection 最惠价中标
qualifications-based selection 资质评审
lump sum 包干
cost plus fee 成本加酬金
guaranteed maximum price 最高成本限额
unit price 单价

procurement [prəˈkjuəmənt] n. 采购
milestone [ˈmaɪlstəʊn] n.
里程碑，重要阶段
diary [ˈdaɪəri] n. 日记，日志
log [lɒg] n. 日志，记录
daily field report 每日现场报告
mediation [ˌmiːdiˈeɪʃn] n. 调停，调解
minitrial n. 法庭外审查，小型庭审
arbitration [ˌɑːbɪˈtreɪʃn] n. 仲裁，公断

Notes 注释

[1] Construction project management (CM) is a professional service that uses specialized, project management techniques to oversee the planning, design, and construction of a project, from its beginning to its end. The purpose of CM is to control a project's time, cost and quality. CM is compatible with all project delivery systems, including design-bid-build, design-build, CM At-Risk and Public Private Partnerships.

建设项目管理（CM）是一种专业服务，它使用专门的项目管理技术来监督项目从项目开始到结束的规划、设计和施工。CM 的目的是控制项目的时间、成本和质量。CM 与所有项目交付系统兼容，包括设计—投标—建造、设计构建、CM 风险管理和公私合作伙伴关系。

[2] A bid is given to the owner by construction managers that are willing to complete their construction project. A bid tells the owner how much money they should expect to pay the construction management company in order for them to complete the project.

——Open bid: An open bid is used for public projects. Any and all contractors are allowed to submit their bid due to public advertising.

——Closed bid: A closed bid is used for private projects. A selection of contractors is sent an invitation for bid so only they can submit a bid for the specified project.

投标由愿意完成施工项目的施工经理向业主提出。投标告诉业主为了完成项目，他们应该向施工管理公司支付多少钱。

——公开招标：公开招标用于公共项目。任何和所有承包商都可以通过公开广告而提交投标书。

——非公开投标：非公开投标用于私人项目。一些承包商会收到招标邀请，所以只有他们可以提交指定项目的投标。

[3] The procurement stage is when labor, materials and equipment needed to complete the project are purchased. This can be done by the general contractor if the company does all their own construction work. If the contractor does not do their own work, they obtain it through subcontractors. Subcontractors are contractors who specialize in one particular aspect of the construction work such as concrete, welding, glass, or carpentry. Subcontractors are hired the same way a general contractor would be, which is through the bidding process. Purchase orders are also part of the procurement stage.

采购阶段是指采购项目所需的劳动力、材料和设备的阶段。如果公司自己做所有的施工工作，这可以由总承包商来做。如果承包商不完成所有的施工，他们通过分包商获得部分服务。分包商是专门从事混凝土、焊接、玻璃或木工等施工工作的某一方面的承包商。分包商的雇用方式与总承包商相同,通过招标程序进行。采购订单也是采购阶段的一部分。

[4] When construction vehicles are driving around a site or moving earth, a lot of dust is created, especially during the dryer months. This may cause disruption for surrounding businesses or homes. A popular method of dust control is to have a water truck driving through the site spraying water on the dry dirt to minimize the movement of dust within and out of the construction site. When water is introduced mud is created. This mud sticks to the tires of the construction vehicles and is often lead out to the surrounding roads. A good practice is to have a street sweeper clean the roads at least once a day to minimize dirty road conditions.

当施工车辆行驶在现场或进行土方工程时,会产生大量的灰尘,尤其是在干燥的月份。这可能会对周围的企业或家庭造成干扰。一种常用的防尘方法是通过洒水车在工地上洒水,以减少灰尘在施工现场内外的扩散。当水被引进时,泥浆就产生了。泥浆粘在施工车辆的轮胎上,经常会洒落到周围的道路上。一个很好的做法是,让一名清洁工至少每天清扫一次道路,以减少道路的肮脏状况。

[5] ——Mediation: Mediation uses a third party mediator to resolve any disputes. The mediator helps both disputing parties to come to a mutual agreement. This process ensures that no attorneys become involved in the dispute and is less time-consuming.

——Minitrial: A minitrial takes more time and money than a mediation. The minitrial takes place in an informal setting and involves some type of advisor or attorney that must be paid. The disputing parties may come to an agreement or the third party advisor may offer their advice. The agreement is nonbinding and can be broken.

——Arbitration: Arbitration is the most costly and time-consuming way to resolve a dispute.

——调解:调解使用第三方调解人解决任何争议。调解人帮助双方达成一致意见。这一过程确保没有律师参与争议,而且省时省力。

——小型庭审:小型庭审要比调解耗费更多的时间和金钱。小型庭审发生在非正式场合,对某些类型的顾问或律师需要支付费用。争议各方可以达成协议,也可以由第三方顾问提出意见。协议是无约束力的,可以违反。

——仲裁:仲裁是解决争议最费钱、最费时的方法。

[6] The construction industry typically includes three parties: an owner, a licensed designer (architect or engineer) and a builder (usually known as a general contractor). There are traditionally two contracts between these parties as they work

together to plan, design and construct the project. The first contract is the owner-designer contract, which involves planning, design, and construction contract administration. The second contract is the owner-contractor contract, which involves construction. An indirect third-party relationship exists between the designer and the contractor, due to these two contracts.

建筑业通常包括三方：业主、注册设计师（建筑师或工程师）和建造商（通常称为总承包商）。传统上，这三方之间会签订两份合同，他们一起计划、设计和完成施工项目。第一份合同是业主—设计师合同，涉及规划、设计和施工合同管理。第二份合同是业主—承包商合同，涉及施工。由于这两份合同，设计师和承包商之间存在间接的第三方关系。

Reference 课文来源

[1] https://en.wikipedia.org/wiki/Construction_management

Unit 12　Timber Engineering

> ▶ 学习任务　木建筑
> ▶ 教学时间　2 学时
> ▶ 学习目标　了解木结构的特点和现场安装的技术要点。

Text 课文

Wood construction (shown in Figure 12-1) significant advantages in such areas as energy savings and total cost at both household and national levels, the inherent resistance to earthquakes which have devastated parts of China over the last century, and the major contributions to China's environmental objectives, including potentially substantial reductions of CO_2 emissions.[1] It addresses limitations of wood used in construction, but also and importantly, debunks misunderstandings about building codes, costs, fire safety, durability, land use, and deforestation.

Figure 12-1　Wood Construction

Building with wood has a Chinese tradition and is proven over the centuries. Wood construction offers solutions for China, including seismic performance and energy conservation. It is popular for single and multi-family housing and suitable for commercial and public buildings.

It is appropriate for medium-rise buildings which address China's wider housing needs. It can be used in combination with concrete structures to improve new and existing buildings. Structural glulam, with its strong aesthetic appeal, is ideal for large span construction. Using wood in building structures is nothing new-China has been building with wood for thousands of years. It has been used as a building material throughout the ages wherever forests grow. And today, the international timber trade provides countries which do not have extensive forest resources with wood from sustainable and certified forestry to build with. Experience, research and product development have resulted in a range of effective building codes and standards. Building with wood is becoming increasingly popular as countries around the world seek more sustainable construction; already 70 percent of the housing constructed in the developed world use wood frame.

12.1 On-site Construction

The traditional way to construct wood frame buildings in North America is on-site, particularly when there is labour availability. In Europe, wood frame assemblies are typically pre-fabricated. Engineered wood construction, such as glulam, is most commonly erected piece by piece on-site. In China, almost all wood buildings are currently constructed on-site (shown in Figure 12-2). Building materials and structural components are freighted to the building site and the various assemblies—walls, floors, etc.—are framed on-site.[2] The method requires organization and planning on the building site and measures must be taken to avoid moisture damage to materials. On-site construction relies on a skilled work force and, while much faster than using other materials, is slower than using prefabricated elements.

Figure 12-2　On-site Construction

On-site construction does not require the initial capital costs for plant and machinery, nor the need to maintain capacity utilization. It is particularly appropriate where housing volumes are not large, where labour is reasonably priced and plentiful, and where flexibility and low overheads are important. While more capital-intensive, off-site pre-fabrication has the benefit of controlled factory conditions, less dependence on on-site labour and faster construction times. In the case of wood frame construction, only a few days on the building site are needed to assemble a water-tight structure, complete with roof. The panels can be pre-fabricated with insulation, windows and doors. Entire units can even be made complete with electricity, water and waste pipes, kitchens and wet rooms, floors and papered walls.

12.2 Pre-fabricated Wood Frame

Although concrete and steel are more common construction materials in China, the government is looking at different solutions, like wood building, as part of its sustainability strategy.

At present, wood frame is used for single family and multi-family homes of two or three storeys in China. Wood members form a structural framework which is sheathed with structural wood panels. Foundations are generally concrete. The floor above can be either wood or a concrete slab and forms the platform for the next storey. Roof and wall insulation and water-proof membranes provide energy-efficiency and protection from moisture. Interiors are usually dry-lined with fire-resistant gypsum board, and many different materials can be used for external cladding. Because the structure has multiple wood members, panels, fasteners and connectors, loads can be carried through a number of alternative pathways. As a result, wood frame buildings are highly resistant to sudden failure in earthquakes or high winds.

Pre-fabricated components, as shown in Figure 12-3, are relatively light and can be erected at heights of several storeys using simple lifting equipment, such as the cranes on the trucks that deliver components to site. Components may need protection against the elements to prevent dampness. The extent of pre-fabrication varies widely between countries and companies, depending on economic factors.[3] It does require an up-front investment in plant and equipment which could impose an uncompetitive burden. While this is essentially the case in China at present, over the longer term, pre-fabrication may prove advantageous.

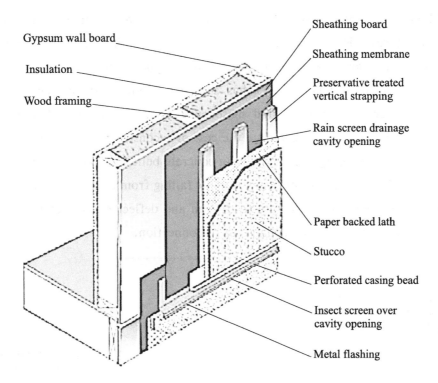

Figure 12-3 Pre-fabricated Components

Wood frame construction is already being used for housing in China, single family dwellings in the suburbs of cities such as Shanghai and Beijing, to low-cost rural developments where land availability is not a problem. They have proved cost-competitive and perform well in comparison with concrete and steel frame housing. But much more can be done. Wood construction is the solution to other building requirements in China as well. These include medium-density multi-storey apartments, small commercial and office buildings, schools, medical clinics, nursing homes, universities and research centres, sports arenas and other recreational facilities.

Wood can make a contribution to solving China's housing shortages through high density multi-family solutions. While these can take the form of two or three-storey apartment blocks, the future lies in the higher-rise buildings which are well-proven in Europe and North America.

They have gained popularity in these regions because of lower building costs, wood's suitability for highly efficient industrial building methods, better energy-efficiency, better seismic performance and a growing environmental awareness. And, because of their low weight, multistorey wood buildings can be constructed without the need for extensive pile foundations. This makes it possible to develop sites which would previously have been impractical.

In China, as of 2009, existing fire codes do not allow wood frame apartment blocks of four or more storeys. However, this may be an option for the future, as these codes are often under review and the scientific experience supports more storeys. Wood frame construction (shown in Figure 12-4) has superior seismic performance. Wood frame buildings are safer than concrete and masonry buildings in areas with a high risk of earthquakes. They save lives and reduce the cost of reconstruction. Wood is strong, light and flexible. Wood buildings weigh less than concrete buildings. This reduces loads on the structure, as well as the danger of heavy weights falling from above. The flexibility of the wood components allows the structure to deform and deflect momentarily in response to seismic forces without breakage, collapse or disconnection.

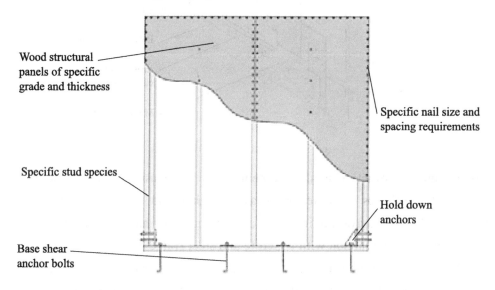

Figure 12-4　Engineered Shear Wall

Uplift and lateral loads are shared by the many wood members that make up the framework, the wood structural panels fastened to them, and the thousands of fasteners and connectors which tie the components together. This structural redundancy is stronger than predicted by conventional engineering analysis.

Additional measures can be taken in areas of greatest risk. In areas such as Sichuan, where severe earthquakes are likely, the structural design of a standard wood frame can be enhanced simply and inexpensively. Additional measures include braced walls, reinforced-connections between foundation and floor, and walls to roof, as well as steel rod tie-downs that clamp the top wall to the foundation.

12.3 Durability of Timber Structure

Appropriate design, material selection, construction, and maintenance will ensure that wood is safe from the decay and mould associated with exposure to excessive moisture, as well as from the termite infestations found in the southern China regions. Durability depends on protecting wood from excessive moisture. Building envelopes should be designed to prevent water vapour condensation within envelope cavities and to allow any dampness to dry out. In wet climates, more steeply sloped roofs, large overhangs and rain screens can be used. As with any building system, the building envelope must be sealed against rain penetration around windows, doors, and other exterior wall openings, including roof penetrations and balconies.[4] In areas with termite hazard, effective prevention and control can be achieved with appropriate design and construction practices. In recent years, multiple lines of defense have been developed and integrated into modern wood construction to ensure moisture and termite resistance.

Exterior wood products used for decks and other landscaping projects are either made from naturally durable wood species, such as the heartwood of China fir, Western red cedar and yellow cedar, or pressure treated with chemical preservatives. Strict environmental and health regulations ensure these chemicals are benign to humans but resistant to insects and fungi. Good design, workmanship, and maintenance are also critical for prolonging the service life of outdoor wood products.

12.4 Codes for Timber Structure

Codes and standards for wood buildings have been evolving quickly so as to encourage new opportunities for different applications. For example, proposed revisions to GB 50016 (national fire code) will, once approved, enable new forms of hybrid construction (combining wood with concrete and/or steel in one structure) and the use of infill wood walls in mid-rise concrete structures, glulam structures. In fact, this should expand the use of wood in construction and contribute to improvements overall in the Chinese building sector.

The national code, Technical Code for Partitions with Timber Framework GB/T 50361—2005, was issued by Ministry of Housing and Urban Rural Development (MHURD) in 2005. It is the first wall code for wood in China. The code is applicable to residential, office and industry buildings. It covers basic requirements for materials, design, production and maintenance of partitions with timber framework, specifying the approval and acceptance process. This code is taken as an example by the Chinese government for applying more energy-efficient building systems and solutions.

The technical code for partitions with timber framework provides guidance and requirements for the installation within existing concrete structures of exterior (up to six stories) and interior (up to 18 stories) infill wood walls, opening up new applications for the use of wood construction in taller buildings.

Glossary 词汇表

timber ['tɪmbə(r)] n. 木材，木料
emission [i'mɪʃn] n. 排放
debunk [ˌdiː'bʌŋk] vt. 揭穿真相，暴露
deforestation [ˌdiːˌfɒrɪ'steɪʃn] n. 森林采伐
forestry ['fɒrɪstri] n. 林业
on-site n. 现场

freight [freɪt] n. 货运；vt. 运输
pre-fabricated n. 预制
sustainability n. 可持续性
uplift ['ʌplɪft] vt. 举起，振作；
    vi. 上升，升起
Ministry of Housing and Urban Rural Development 住房和城乡发展部

Notes 注释

[1] Wood construction (shown in Figure 12-1) has significant advantages in such areas as energy savings and total cost at both household and national levels, the inherent resistance to earthquakes which have devastated parts of China over the last century, and the major contributions to China's environmental objectives, including potentially substantial reductions of CO_2 emissions.

木建筑具有很多重大优势，有利于节能和控制家庭和国家层面的总成本；具备固有的抗地震能力，上世纪中国部分地区曾遭受地震破坏；它对中国的环境目标也有重大贡献，包括可能大幅度减少二氧化碳排放。

[2] The traditional way to construct wood frame buildings in North America is on-site, particularly when there is labour availability. In Europe, wood frame assemblies are typically pre-fabricated. Engineered wood construction, such as glulam, is most commonly erected piece by piece on-site. In China, almost all wood buildings are currently constructed on-site (shown in Figure 12-2). Building materials and structural components are freighted to the building site and the various assemblies—walls,

floors, etc. are framed on-site.

在北美，建造木结构房屋的传统方法是现场施工，特别是在有劳动力的情况下。在欧洲，木框架组件通常是预制的。工程木结构，如胶合木，最常见的是在现场搭建。在中国，几乎所有的木建筑物都是现场建造的（图12-2）。建筑材料和结构部件运送到建筑工地，各种组件如墙壁、地板等都在现场构筑。

[3] Pre-fabricated components, as shown in Figure 12-3, are relatively light and can be erected at heights of several storeys using simple lifting equipment, such as the cranes on the trucks that deliver components to site. Components may need protection against the elements to prevent dampness. The extent of pre-fabrication varies widely between countries and companies, depending on economic factors.

预制部件（图12-3）比较轻，可以用简单的起重设备如卡车上的起重机将部件送到现场，在几层楼高的地方架设。组件可能需要保护构件以防止潮湿。取决于经济因素，国家和公司之间的预制程度差异很大。

[4] Durability depends on protecting wood from excessive moisture. Building envelopes should be designed to prevent water vapour condensation within envelope cavities and to allow any dampness to dry out. In wet climates, more steeply sloped roofs, large overhangs and rain screens can be used. As with any building system, the building envelope must be sealed against rain penetration around windows, doors, and other exterior wall openings, including roof penetrations and balconies.

耐久性取决于保护木材免受过度潮湿。建筑围护结构应设计成防止水蒸气在封闭空间内凝结，并允许任何湿度的木材变干燥。在潮湿的气候下，可以使用更陡的斜屋顶、大挑檐、雨屏。与任何其他建筑体系一样，木建筑围护结构也必须进行密封以防止雨水渗透窗户、门和其他外墙开口，包括屋顶渗漏和阳台。

Reference 课文来源

[1] https://en.wikipedia.org/wiki/Framing_(construction)

Unit 13 Prefabricated Building

> ▸ 学习任务 预制装配结构
> ▸ 教学时间 2 学时
> ▸ 学习目标 了解预制装配结构的技术要求和发展现状。

Text 课文

A prefabricated building, informally a prefab, is a building that is manufactured and constructed using prefabrication. It consists of factory-made components or units that are transported and assembled on-site to form the complete building.

"Prefabricated" may refer to buildings built in components (e.g. panels), modules (modular homes) or transportable sections (manufactured homes), and may also be used to refer to mobile homes, i.e., houses on wheels. Although similar, the methods and design of the three vary widely.[1] There are two-level home plans, as well as custom home plans. There are considerable differences in the construction types.

Modular homes are created in sections, and then transported to the home site for construction and installation. These are typically installed and treated like a regular house, for financing, appraisal and construction purposes, and are usually the most expensive of the three. Although the sections of the house are prefabricated, the sections, or modules, are put together at the construction much like a typical home. Manufactured and mobile houses are rated as personal property and depreciate over time.

Manufactured homes are built onto steel beams, and are transported in complete sections to the home site, where they are assembled. Mobile homes are built on wheels, that can be moved. Mobile homes and manufactured homes can be placed in mobile home parks, and manufactured homes can also be placed on private land, providing the land is zoned for manufactured homes.

13.1 History

The world's first prefabricated, pre-cast panelled apartment blocks were pioneered in Liverpool. A process was invented by city engineer John Alexander Brodie, whose inventive genius also had him inventing the football goal net. The tram stables at Walton in Liverpool followed in 1906. The idea was not extensively adopted in Britain, however was widely adopted elsewhere, particularly in Eastern Europe.

Prefabricated homes were produced during the Gold Rush in the United States, when kits were produced to enable Californian prospectors to quickly construct accommodation. Homes were available in kit form by mail order in the United States in 1908.

Prefabricated housing was popular during the Second World War due to the need for mass accommodation for military personnel. The proliferation of prefabricated housing across the country was a result of the Burt Committee and the Housing (Temporary Accommodation) Act 1944. Almost 160,000 had been built in the UK by 1948 at a cost of close to £216 million. The largest single prefab estate in Britain was at Belle Vale (South Liverpool), where more than 1,100 were built after World War II. The estate was demolished in the 1960s amid much controversy as the prefabs were very popular with residents at the time.

13.2 Prefabricated Houses

Prefabs were aimed at families, and typically had an entrance hall, two bedrooms (parents and children), a bathroom (a room with a bath) —which was a novel innovation for many Britons at that time, a separate toilet, a living room and an equipped (not fitted in the modern sense) kitchen. Construction materials included steel, aluminium, timber or asbestos, depending on the type of dwelling.[2] The aluminium Type B2 prefab was produced as four pre-assembled sections which could be transported by lorry anywhere in the country.

Many buildings were designed with a five-ten year life span, but have far exceeded this, with a number surviving today. In 2002, for example, the city of Bristol still had residents living in 700 examples. Many UK councils have been in the process of demolishing the last surviving examples of Second World War prefabs in order to comply with the British government's Decent Homes Standard, which came into effect in 2010. There has, however, been a recent revival in prefabricated methods of construction in order to compensate for the United Kingdom's current housing shortage.

Architects are incorporating modern designs into the prefabricated houses of today. Prefab housing should no longer be compared to a mobile home in terms of appearance, but to that of a complex modernist design. There has also been an increase in the use of "green" materials in the construction of these prefab houses. Consumers can easily select between different environmentally friendly finishes and wall systems. Since these homes are built in parts, it is easy for a home owner to add additional rooms or even solar panels to the roofs. Many prefab houses, as shown in Figure 13-1, can be customized to the client's specific location and climate, making prefab homes much more flexible and modern than before.

There is a zeitgeist in architectural circles and the spirit of the age favors the small carbon footprint of "prefab." Eminent amongst the new breed of off the shelf luxury modernist products is the perrinepod, which has found favor worldwide for its green credentials and three-day build time.

Figure 13-1　Prefabricated Family Home in Princeton

Prefab homes, as shown in Figure 13-2, are becoming popular in Europe, Canada and United States as they are relatively cheap when compared to many existing homes on the market. The 2007 finance crisis has however deflated the cost of housing in North America and Europe, so not all prefab homes should be assumed to be cheaper than existing housing.

Figure 13-2　Fibre-cement Cladding on the Entry Facade

　　Modern architects are experimenting with prefabrication as a means to deliver well-designed and mass-produced modern homes. Modern architecture forgoes referential decoration and instead features clean lines and open floor plans.

　　Because of the design simplifications modern architecture provides (coupled with the cost savings that tend to go with design simplification) many in the manufactured housing sector generally feel that modern architecture designs are better suited for prefab home construction.

13.3　Prefabricated Workshop

　　Prefabricated workshop (shown in Figure 13-3—Figure 13-5) is widely applied to engineering practice and has the following good properties:

　　—Easy to assemble and disassemble with simple and common tools.

　　—Heat and sound insulation, water and fire proofing.

　　—Attractive appearance. the whole structure is handsome with adopting the color coated steel sheet.

　　—Environmentally friendly and economical. Reasonable design makes it reusable. The reusable character makes it environmentally friendly and economical.

　　—Cost efficient. First class materials, reasonable price, once and for all investment, low requirements for base and short completion time make it cost efficient.

　　—Light weight, convenient for shipment and transportation.

—Components of the house can be used repeatedly.

—Beautiful appearance, various colors and shapes for outer and inner roof panel and wall panel.

—Various designs available, customized designs acceptable.

—Wide range of application, could be used as offices, command posts, sentry boxes, dormitories, shops, Kiosks and so on.[3]

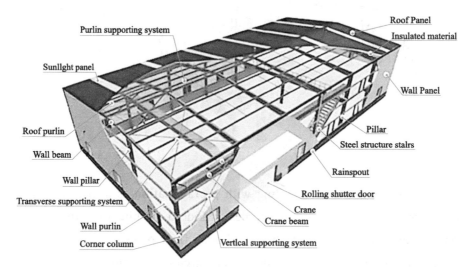

Figure 13-3　Structure of Prefabricated Workshop

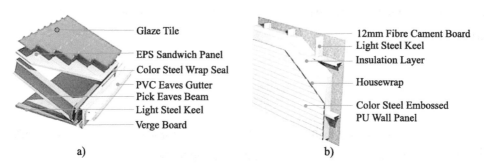

Figure 13-4　Roof and Wall System
a) roof system　b) wall system

Figure 13-5　Connection Point

Glossary 词汇表

tram [træm] n. 电车轨道
asbestos [æsˈbestəs] n. 石棉;adj. 石棉制的,含石棉的
dwelling [ˈdwelɪŋ] n. 居住,住处,寓所
prefab [ˈpriːfæb] adj. 预制的;n. 活动房屋
zeitgeist [ˈzaɪtɡaɪst] n. (尤指文学、哲学和政治中表现出的)时代精神,时代思潮
eminent [ˈemɪnənt] adj. 显赫的,明显的,突出的
credential [krəˈdenʃl] n. 文凭,信任状;v. 提供证明书

Notes 注释

[1] "Prefabricated" may refer to buildings built in components (e.g. panels), modules (modular homes) or transportable sections (manufactured homes), and may also be used to refer to mobile homes, i.e., houses on wheels. Although similar, the methods and design of the three vary widely.

"预制"可以是指在构件(如面板)、模块(模块式住宅)或可移动部分(制造的房屋)中建造的建筑物,也可以用来指移动房屋,即在轮子上的房屋。虽然相似,但三者的方法和设计大相径庭。

[2] Prefabs were aimed at families, and typically had an entrance hall, two bedrooms (parents and children), a bathroom (a room with a bath) —which was a novel innovation for many Britons at that time, a separate toilet, a living room and an equipped (not fitted in the modern sense) kitchen. Construction materials included steel, aluminum, timber or asbestos, depending on the type of dwelling.

预制屋针对的是家庭,通常有一个入口大厅,两间卧室(父母和孩子),一间浴室(一间带浴缸的房间)间(带浴室),这在当时对许多英国人来说是一个新奇的创新,一个单独的卫生间,一个客厅和一个装备化的(不是现代意义上的)厨房。建筑材料包括钢材、铝、木材或石棉,具体取决于住宅的类型。

[3] Prefabricated workshop is widely applied to engineering practice and has the following good properties:

—Easy to assemble and disassemble with simple and common tools.

—Heat and sound insulation, water and fire proofing.

—Attractive appearance. The whole structure is handsome with adopting the color coated steel sheet.

—Environmentally friendly and economical. Reasonable design makes it reusable. The reusable character makes it environmentally friendly and economical.

—Cost efficient. First class materials, reasonable price, once and for all investment, low requirements for base and short completion time make it cost efficient.

—Light weight, convenient for shipment and transportation.

—Components of the house can be used repeatedly.

—Beautiful appearance, various colors and shapes for outer and inner roof panel and wall panel.

—Various designs available, customized designs acceptable.

—Wide range of application, could be used as offices, command posts, sentry boxes, dormitories, shops, Kiosks and so on.

预制车间广泛应用于工程实践，具有以下优良性能：

——使用简单和常见的工具，易于组装和拆卸。

——隔热、隔音，防水、防火。

——造型美观，采用彩钢板，整体结构美观。

——环保和经济。合理的设计使其具有可重复使用性，可重复使用的特点使它环保、经济。

——成本效益。一流的材料，合理的价格，一劳永逸的投资，基础要求低，完成时间短，使其具有成本效益。

——重量轻，便于运输。

——房子的部件可以反复使用。

——外观美观，内、外屋面板及墙面板均有多种颜色和形状。

——有各种设计可供选择，可接受定制设计。

——用途广泛，可作办公室、指挥所、岗亭、宿舍、商店、报刊亭等。

References 课文来源

[1] http://www.authorstream.com/Presentation/greenrpanel-2781278-manufactured-homes/
[2] https://en.wikipedia.org/wiki/Prefabricated_building
[3] https://www.alibaba.com/product-detail/Long-life-stainless-steel-low-cost_60054448082.html

Unit 14 Building Information Modeling

> ▸ 学习任务　建筑信息模型
> ▸ 教学时间　2 学时
> ▸ 学习目标　了解以 BIM 技术为代表的建筑信息模型在土木工程中的应用。

Text 课文

Buildings and public infrastructure have a major significance to society as they support the life and social activities of our citizens. Additionally, they have great environmental impact, consuming large amounts of energy and other resources. Global construction spending is increasing rapidly and is forecast to reach 13.2% of world GDP by 2020. The exponential complexity of the contemporary city exceeds the capacity of traditional modeling and visualization methods and requires more sophisticated technologies that can enable professionals to collaborate across disciplinary boundaries and diverse data sets.

The construction industry has been moving towards the industrialization of construction, based on lean principles and practices, to refine the end-to-end design and construction process. Buildings have a major significance to society because they are integral parts of the basic infrastructure that supports the life and social activities of our citizens. Additionally, they have great environmental impact, consuming large amounts of water, energy, and other resources. The building industry is expected not only to cope with tightening government building codes in different countries and ensure compliance, but also to pursue progress through voluntary efforts, working toward such goals as reducing energy consumption to zero, expanding the use of recyclable construction materials, and developing a new delivery mechanism to better respond to owner demands.

In recent years, the building industry has been accelerating the process of industrialization, learning from the highly efficient production systems employed by the automobile and aircraft industries. For example, the industry has been expanding the use of pre-fabricated components from off-site factories to improve efficiency of on-site activities, and has introduced just-in-time delivery systems to use storage spaces more efficiently, improving cash flow at the same time.[1]

14.1 Building Information Modeling

Building Information Modeling (BIM) is a set of processes and methods to organize and structure geometry and data related to the design and construction of a building or an infrastructure. BIM-related software provides project stakeholders with a virtual equivalent of a structure in the form of a 3D digital model. Every component in the digital mock-up is the sum of its geometry and other related attributes including supplier information, part reference number, and cost. All this information exists in one place, providing a thorough and comprehensive view that facilitates the work of the various disciplines participating in a project.[2] Through BIM, professionals can efficiently plan, design, build and maintain a building or an infrastructure because all information is available to project members at all times throughout the structure's lifecycle. Different disciplines no longer work in silos but in a concerted way, contributing their expertise and trying out new ideas in real time to complete the virtual mock-up. All information is capitalized in the CAD model providing traceability throughout the building's lifecycle. With this digital equivalent, stakeholders can perform simulations to detect interferences and construction issues in the early design stages when modifications least impact budgets and schedules. They can simulate the various functions of a building such as the heating system, environmental controls, and emergency procedures providing valuable insights into maintenance and sustainability constraints. As environmental regulations grow more complex, building professionals can benefit from BIM to improve energy-efficiency and reduce waste. Building Information Modeling enhances information transparency thereby promoting a more streamlined design and construction process.

The term "building model" (in the sense of BIM as used today) was first used in papers in the mid-1980s. However, the terms "Building Information Model" and "Building Information Modeling" (including the acronym "BIM") did not become popularly used until some 10 years later. In 2002, Autodesk released a white paper entitled "Building Information Modeling," and other software vendors also started to assert their involvement in the field.

Traditional building design was largely reliant upon two-dimensional technical drawings (plans, elevations, sections, etc.). Building information modeling extends this beyond 3D, augmenting the three primary spatial dimensions (width, height and depth) with time as the fourth dimension (4D) and cost as the fifth (5D). BIM therefore covers more than just geometry. It also covers spatial relationships, light analysis, geographic information, and quantities and properties of building components.

BIM involves representing a design as combinations of "objects"—vague and undefined, generic or product-specific, solid shapes or void-space oriented (like the shape of a room), that carry their geometry, relations and attributes. BIM design tools allow extraction of different views from a building model for drawing production and other uses. These different views are automatically consistent, being based on a single definition of each object instance. BIM software also defines objects parametrically; that is, the objects are defined as parameters and relations to other objects, so that if a related object is amended, dependent ones will automatically also change. Each model element can carry attributes for selecting and ordering them automatically, providing cost estimates as well as material tracking and ordering.

For the professionals involved in a project, BIM enables a virtual information model to be handed from the design team (architects, landscape architects, surveyors, civil, structural and building services engineers, etc.) to the main contractor and subcontractors and then on to the owner/operator; each professional adds discipline-specific data to the single shared model. This reduces information losses that traditionally occurred when a new team takes "ownership" of the project, and provides more extensive information to owners of complex structures.

14.2 Application of BIM

Use of BIM goes beyond the planning and design phase of the project, extending throughout the building life cycle, supporting processes including cost management, construction management, project management and facility operation.

BIM can be utilized for successful construction, lifecycle and long term facility management. BIM data is applied to demonstrate the entire building life cycle. Quantities and properties of materials are extracted without any difficulties. The scope of works is defined effortlessly. Moreover, assemblies and sequences are presented in a comparative scale to each other and relative to the entire project.[3]

BIM contains geometry, spatial relationships, light analysis, geographic information, quantities and properties of building components. Building Information Modeling is digital representation of physical (actual parts and pieces being used in the construction process) and functional characteristics of a facility producing an allocated knowledge source of information about the facility. Thus BIM outlines a trustworthy basis for decisions throughout its life cycle, from earliest conception to demolition.

14.2.1 Management of building information models

Building information models span the whole concept-to-occupation time-span. To ensure efficient management of information processes throughout this span, a BIM manager (also sometimes defined as a virtual design-to-construction, VDC, project manager—VDCPM) might be appointed. The BIM manager is retained by a design build team on the client's behalf from the pre-design phase onwards to develop and to track the object-oriented BIM against predicted and measured performance objectives, supporting multi-disciplinary building information models that drive analysis, schedules, take-off and logistics. Companies are also now considering developing BIMs in various levels of detail, since depending on the application of BIM, more or less detail is needed, and there is varying modeling effort associated with generating building information models at different levels of detail.

Just after the launching of BIM, the contractors quickly adopt the technique to get best result in their business. BIM Outsourcing helps the contractors to get the winning bid. With the help of BIM, the contractors can visualize the project far before the groundbreaking. They can check the interference. A proper and planning can be done by the building information modeling. It is quite easy to understand "what if" situation. In BIM Outsourcing the contractor can get the high quality, time saving service in an affordable price.

The owner should embrace BIM to reap the benefit of a well-organized and extremely coordinated method of project delivery. Project schedules are also curtailed and overall project costs are decreased. A BIM model contains huge database of vital information and the owner can utilize this information for analysis and decision making. The database is repetitively updated and filtered during the life cycle of the model. Besides modification can be made at a single location in spite of how many times the data is presented or distributed. The recurrence of requesting, parsing, assembling and transmitting information among team members at the different arenas of a model's life is significantly decreased or removed. This leads to improved communication between team members and more apt reply to owner concerns or suggestions. BIM generates coordinated construction documents which can enhance the efficiency of the design and construction procedure. The owner can gain huge cost saving as the project scheduled is accelerated to a great extent. The owner can apply BIM at pre-design phase and expand it during the design, bidding, construction, occupancy and finally decommissioning of the building.

BIM plays a key role in allocating data for analysis and simulation. BIM makes it possible to develop the amount and types of project analysis. This may vary from early investigation of building energy efficiency throughout conceptual massing, to structural / mechanical / electrical load analysis all through design, quantity take-offs throughout the project bidding, shop drawing review/fabrication all through construction, or even facility asset management throughout occupancy. The owner obtains more value by allocating different analyses to be executed with the transmission of data. This is applicable where numerous iterations of an analysis are replicated and purified at different phases of the BIM lifecycle.

14.2.2 BIM in construction management

BIM or Building Information with its real-time and three-dimensional features, serve as a construction productivity tool to increase productivity in the design and construction phases. Many large construction companies are experimenting with BIM to produce cost and schedule savings (shown in Figure 14-1).

Participants in the building process is constantly challenged to deliver successful projects despite tight budgets, limited manpower, accelerated schedules, and limited or conflicting information. The significant disciplines such as architectural, structural and Mechanical, Electrical & Plumbing (MEP) designs should be well coordinated, as two things can't take place at the same place and time. Building Information Modeling aids in collision detection at the initial stage, identifying the exact location of discrepancies.[4]

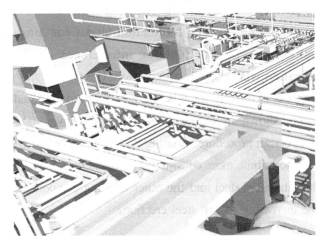

Figure 14-1　Installation with Building Information Model

The BIM concept envisages virtual construction of a facility prior to its actual physical construction, in order to reduce uncertainty, improve safety, work out problems, and simulate and analyze potential impacts. Sub-contractors from every trade can input critical information into the model before beginning construction, with opportunities to pre-fabricate or pre-assemble some systems off-site. Waste can be minimized on-site and products delivered on a just-in-time basis rather than being stock-piled on-site.

Quantities and shared properties of materials can be extracted easily. Scopes of work can be isolated and defined. Systems, assemblies and sequences can be shown in a relative scale with the entire facility or group of facilities. BIM also prevents errors by enabling conflict or "clash detection" whereby the computer model visually highlights to the team where parts of the building (e. g. structural frame and building services pipes or ducts) may wrongly intersect.

BIM is very useful for the digital fabrication of structural steel. BIM can model entire steelwork structures in 3D along with last notch, plate and bolt to facilitate the fabricators in reviewing a 3D steel structural model right from the fabricated components. All detailed information as well as the drawings, material list and other data are automatically generated from the model. The various data can then be transmitted to production planning and automation systems. Such automated routine tasks permit the fabricator more time to focus on the principle precast design and detailing decisions.

A purpose-built building information model can nourish the manufacturing process as the design model contains all the geometry related with structural steel. The design information is exported to a CIS/2 file (an industry standard data format for replacing steel information) to be used again in a steel detailing application.

Once the steel detailing is completed, the design team or contractor utilize the fabrication model for 4D modeling and clash detection with other building disciplines and models (MEP, architectural, etc.). The fabrication model contains more detail than the structural model, and consequently helpful for interference checking, particularly for building types or application where space is extremely tight.

Applying the design model directly for fabrication will produce a usual feedback loop among fabricators and designers (shown in Figure 14-2). It will convey fabrication considerations forward into the building design procedure. Sharing the design model with fabricators for bidding will curtail the bid cycle and initiate more consistent bids based on reliable steel tonnages. And the coordination among the fabricated steel and the other building components will decrease the on-site issues and lower the increasing cost of steel erection.

Figure 14-2 Construction with Building Information Model

An integrated BIM model empowers fabricators to manage not only the fabrication phase of the project but also other verticals of the project as well as modeling, fabricated parts design and elaborating, contract management, financial evaluation, and supply chain management, along with keeping track of updated manufacturing data.

BIM can be utilized for construction visualization and results in considerable cost savings, from design and construction through to maintenance. BIM eliminates a lot of risks and issues easier and earlier before actual construction took place. As a result, the construction process gets shortened and construction takes places more efficiently.

BIM model saves time and waste on site as the model provides huge accurate and coordinated information which will lead to fewer errors on site. Thus BIM results in considerable cost savings, from design and construction through to maintenance.

The reduced material waste and aptitude to optimize energy utilization through BIM has also made it a rising option for green building projects.

With its parametric characteristics, BIM tracks the relations among numerous objects from early design through to completion so that if one object changes, the change is automatically coordinated across the project. Each change in design, scheduling or material is immediately accessible to every member of the building team—from engineering and purchasing to plumbing and electrical. For example, BIM analyzes and illustrates how many light fixtures are required to light up a space receiving little sunlight. It works out how a change in wall color matches up against the carpeting that's on order. BIM can also find out how a recommended change will have an effect on subcontractor costs and scheduling.

14.2.3 BIM in facility operation

BIM can bridge the information loss associated with handling a project from design team, to construction team and to building owner/operator, by allowing each group to add to and refer back to all information they acquire during their period of contribution to the BIM model. This can yield benefits to the facility owner or operator.

For example, a building owner may find evidence of a leak in his building. Rather than exploring the physical building, he may turn to the model and see that a water valve is located in the suspect location. He could also have in the model the specific valve size, manufacturer, part number, and any other information ever researched in the past, pending adequate computing power. Such problems were initially addressed by Leite and Akinci when developing a vulnerability representation of facility contents and threats for supporting the identification of vulnerabilities in building emergencies.

Dynamic information about the building, such as sensor measurements and control signals from the building systems, can also be incorporated within BIM software to support analysis of building operation and maintenance.

There have been attempts at creating information models for older, pre-existing facilities. Approaches include referencing key metrics such as the Facility Condition Index (FCI), or using 3D laser-scanning surveys and photogrammetry techniques (both separately or in combination) to capture accurate measurements of the asset that can be used as the basis for a model. Trying to model a building constructed in, say 1927, requires numerous assumptions about design standards, building codes, construction methods, materials, etc., and is therefore more complex than building a model during design.

BIM can potentially offer some benefits for managing stratified cadastral spaces in urban built environments. The first benefit would be enhancing visual communication of interweaved, stacked and complex cadastral spaces for non-specialists. The rich amount of spatial and semantic information about physical structures inside models can aid comprehension of cadastral boundaries, providing an unambiguous delineation of ownership, rights, responsibilities and restrictions.[5] Additionally, using BIM to manage cadastral information could advance current land administration systems from a 2D-based and analogue data environment into a 3D digital, intelligent, interactive and dynamic one. BIM could also unlock value in the cadastral information by forming a bridge between that information and the interactive lifecycle and management of buildings.

Glossary 词汇表

recyclable [ˌriːˈsaɪkləbl] *adj.* 可循环再用的

industrialization [ɪnˌdʌstrɪəlaɪˈzeɪʃn] *n.* 工业化

Building Information Modeling 建筑信息模型

stakeholder [ˈsteɪkhəʊldə(r)] *n.* 股东，利益相关者

attribute [əˈtrɪbjuːt] *vt.* 把……归于；*n.* 属性

traceability [ˌtreɪsəˈbɪləti] *n.* 可追踪性，可追溯

lifecycle [ˈlaɪfˌsaɪkl] *n.* 生命周期

recurrence [rɪˈkʌrəns] *n.* 重现，反复

envisage [ɪnˈvɪzɪdʒ] *vt.* 设想，正视，面对，展望

Mechanical, Electrical, and Plumbing (MEP) 机械、电气、管道

discrepancy [dɪsˈkrepənsi] *n.* 矛盾，不符合（之处）

cadastral [kəˈdæstrəl] *adj.* （有关）土地清册的

semantic [sɪˈmæntɪk] *adj.* 语义的，语义学的

Notes 注释

[1] In recent years, the building industry has been accelerating the process of industrialization, learning from the highly efficient production systems employed by the automobile and aircraft industries. For example, the industry has been expanding the use of pre-fabricated components from off-site factories to improve efficiency of on-site activities, and has introduced just-in-time delivery systems to use storage spaces more efficiently, improving cash flow at the same time.

近年来，建筑业借鉴了汽车、飞机等行业采用的高效生产体系一直处于加速工业化进程中。例如，业界一直在扩大非现场工厂预制部件的使用，以提高现场生产活动的效率，并引入及时交付系统，以便更有效地利用仓储空间，同时也改善了企业的现金流。

[2] Building Information Modeling (BIM) is a set of processes and methods to organize and structure geometry and data related to the design and construction of a building or an infrastructure. BIM-related software provides project stakeholders with a virtual equivalent of a structure in the form of a 3D digital model. Every component in the digital mock-up is the sum of its geometry and other related attributes including supplier information, part reference number, and cost. All this information exists in one place, providing a thorough and comprehensive view that facilitates the work of the various disciplines participating in a project.

建筑信息建模（BIM）是一组过程和方法，用于组织和构造与建筑物或基础设施的设计和施工有关的几何图形和数据。与 BIM 相关的软件为项目利益相关者提供了 3D 数字模型的虚拟结构等价体系。数字样机中的每一个部件都是其几何形状和其他相关属性的总和，包括供应商信息、零件编号和成本。所有这些信息都存在于一个地方，为促进参与项目的各个学科的工作提供了一个彻底而全面的视角。

[3] BIM can be utilized for successful construction, lifecycle and long term facility management. BIM data is applied to demonstrate the entire building life cycle. Quantities and properties of materials are extracted without any difficulties. The scope of works is defined effortlessly. Moreover, assemblies and sequences are presented in a comparative scale to each other and relative to the entire project.

BIM 可以成功地用于建筑、生命周期和长期设施管理。BIM 数据可以应用于整个建筑生命周期的演示。提取材料的数量和性能没有任何困难。BIM 可以毫不费力地定义工程的范围。此外，项目的部件和工序可以呈现彼此相对的比例，并与整个项目相适应。

[4] Participants in the building process is constantly challenged to deliver successful projects despite tight budgets, limited manpower, accelerated schedules, and limited or conflicting information. The significant disciplines such as architectural, structural and Mechanical, Electrical & Plumbing (MEP) designs should be well coordinated, as two things can't take place at the same place and time. Building Information Modeling aids in collision detection at the initial stage, identifying the exact location of discrepancies.

尽管预算紧张、人力有限、进度加快、信息有限或相互冲突，但建造过程中的参与者经常面临着交付成功项目的挑战。因为不能在同一地点和同一时间开展两个学科，如建筑、结构和机械、电气和管道（MEP）设计等重要的学科应协调一致。BIM可以在初始阶段帮助碰撞检测，识别的不一致处的确切位置。

[5] BIM can potentially offer some benefits for managing stratified cadastral spaces in urban built environments. The first benefit would be enhancing visual communication of interweaved, stacked and complex cadastral spaces for non-specialists. The rich amount of spatial and semantic information about physical structures inside models can aid comprehension of cadastral boundaries, providing an unambiguous delineation of ownership, rights, responsibilities and restrictions.

BIM可以为城市建筑环境中的分层地籍空间管理提供一些好处。第一个好处是为非专业人士增强针对交织、堆叠和复杂的地籍空间的视觉传达。模型内部物理结构的空间和语义信息的丰富性可以帮助理解地籍边界，提供明确的所有权、权利、责任和限制的划分。

References 课文来源

[1] https://www.3ds.com/events/all-3dexperience-forum/? woc = %7B%22event%20category%22%3A%5B%22event%20category%2F3dexperience%20forum%22%5D%7D&event + category/3dexperience + forum/&map = 1

[2] https://www.3ds.com/industries/architecture-engineering-construction/what-is-bim/

Appendix

The Contract for Works of Civil Engineering Constrnction
国际土木工程建筑承包合同（中英文对照）

PART I GENERAL CONDITIONS
第一章 总则

DEFINITIONS AND INTERPRETATION
定义和释义

1.1 In the Contract, as hereinafter defined, the following words and expressions shall have the meanings hereby assigned to them, except where the context otherwise requires:

a) "Employer" means the party named in Part II who will employ the Contractor and the legal successors in title to the Employer, but not, except with the consent of the Contractor, any assignee of the Employer.

b) "Contractor" means the person or persons, firm or company whose tender has been accepted by the Employer and includes the Contractor's personal representatives, successors and permitted assigns.

c) "Engineer" means the Engineer designated as such in Part II, or other Engineer appointed from time to time by the Employer and notified in writing to the Contractor to act as Engineer for the purposes of the Contract in place of the Engineer so designated.

d) "Engineer's Representative" means any resident engineer or assistant of the Engineer or any clerk of works appointed from time to time by the Employer or the Engineer to perform the duties set forth in Clause 2 hereof, whose authority shall be notified in writing to the Contractor by the Engineer.

e) "Works" shall include both Permanent Works and Temporary Works.

f) "Contract" means the Conditions of Contract, Specification, Drawings, priced Bill of Quantities, Schedule of Rates and Prices, if any, Tender, Letter of Acceptance and the Contract Agreement, if completed.

g) "Contract Price" means the sum named in the Letter of Acceptance, subject to such additions thereto or deductions therefrom as may be made under the provisions hereinafter contained.

h) "Constructional Plant" means all appliances or things of whatsoever nature required in or about the execution or maintenance of the Works but does not include materials or other things intended to form or forming part of the Permanent Works.

i) "Temporary Works" means all temporary works of every kind required in or about the execution or maintenance of the Works.

j) "Permanent Works" means the permanent works to be executed and maintained in accordance with the Contract.

k) "Specification" means the Specification referred to in the Tender and any modification thereof or addition thereto as may from time to time be furnished or approved in writing by the Engineer.

l) "Drawings" means the drawings referred to in the Specification and any modification of such drawings approved in writing by the Engineer and such other drawings as may from time to time be furnished or approved in writing by the Engineer.

m) "Site" means the land and other places on, under, in or through which the Permanent Works or Temporary Works designed by the Engineer are to be executed and any other lands and places provided by the Employer for working space or any other purpose as may be specifically designated in the Contract as forming part of the Site.

n) "Approved" means approved in writing, including subsequent written confirmation of previous verbal approval and "approval" means approval in writing, including as aforesaid.

1.1 在本合同中，除按上下文另具意义者外，下列词语应解释如下：

a) "业主"指第二章中所指定的雇用承包人的一方或其权利继承人，但不包括业主的受让人，经承包人同意者除外。

b) "承包人"指标书已被业主接受的某个人或某些人、商行或公司，包括其个人代表、继承人和业经认可的受让人。

c) "工程师"指第二章中所指定的工程师，或由业主随时任命且书面通知承包人以代替指定工程师履行合同职责的其他工程师。

d) "工程师代表"指任何常驻工程技术人员、工程师助手，或由业主或工程师随时任命履行本合同第二条规定职责的任何工程现场监督，其权限应由工程师书面通告承包人。

e) "工程"包括永久性工程和临建工程。

f) "合同"指合同条款、技术规范、图纸、标价的建筑工程清单、单价和价格表（如果有），还可指标书、接受证书以及承包协议（如已完成）。

g) "合同价格"指在接受证书中确定的数额，可按本合同以下条款规定增减。

h) "建筑设备"指工程施工和维修中或有关施工和维修所需的全部设备或物品，不论任何性质，但不包括旨在构成或正在构成永久性工程某一部分的材料或其他物品。

i) "临建工程"指工程施工或维修或有关工程施工或维修所需的各种临时工程。

j) "永久性工程"指按照合同将施工和维修的永久工程。

k) "技术规范"指在标书或任何标书更改中提及的规范，或由工程师随时可能增加或书面同意增加的部分。

l) "图纸"指技术规范中规定的图纸，经工程师书面同意对此种图纸所做的任何更改，以及可由工程师随时提供或书面认同的其他图纸。

m) "工地"指工程师设计的永久性或临建工程施工所需的土地及其他场地，包括地面、地下、在之上或通过部分，以及由业主所提供的用作临时储存或其他目的的其他土地或场所，只要能按合同明文规定构成工地的组成部分。

n) "业经认可"指已经经书面认可，包括过后对口头认可的书面确认，"认可"指书面认可，包括上述规定在内。

1.2 Words importing the singular only also include the plural and vice versa where the context requires.

1.2 按合同上下文所需，单数含义的单词也可具有复数的含义，反之一样。

1.3 The headings and marginal notes in these Conditions of Contract shall not be deemed to be part thereof or be taken into consideration in the interpretation or construction thereof or of the Contract.

1.3 合同条款的标题和边注不得视为合同的一部分，不得用于考虑解释条款或合同。

1.4 The word "cost" shall be deemed to include overhead costs whether on or off the Site.

1.4 "费用"一词应视为含工地上或以外发生的间接费用。

ENGINEER AND ENGINEER'S REPRESENTATIVE
工程师及工程师代表

2.1 The Engineer shall carry out such duties in issuing decisions, certificates and orders as are specified in the Contract. In the event of the Engineer being required in terms of his appointment by the Employer to obtain the specific approval of the Employer for the execution of any part of these duties, this shall be set out in Part II of these Conditions.

2.1 工程师必须按合同明文规定，履行做决断、颁发证书和发出指令等职责。如业主签发的工程师任命书中规定其某些职责的履行得经业主专门认可，其要件应在本合同第二部分予以规定。

2.2 The Engineer may from time to time in writing delegate to the Engineer's Representative any of the powers and authorities vested in the Engineer and shall furnish to the Contractor and to the Employer a copy of all such written delegations of powers and authorities. Any written instruction or approval given by the Engineer's Representative to the Contractor within the terms of such delegation, but not otherwise, shall bind the Contractor and the Employer as though it had been given by the Engineer. Provided always as follows:

a) Failure of the Engineer's Representative to disapprove any work or materials shall not prejudice the power of the Engineer thereafter to disapprove such work or materials and to order the pulling down, removal or breaking up thereof.

b) If the Contractor shall be dissatisfied by reason of any decisions of the Engineer's Representative he shall be entitled to refer the matter to the Engineer, who shall thereupon confirm, reverse or vary such decision.

2.2 工程师可随时书面授权其代表代行其任何职权，但必须将所有此种授权书的副本提交给承包人和业主。在授权期间，工程师代表给承包人的任何书面指令或认可（仅限于此）对承包人和业主具有与工程师的指令或认可同样的效力。以下规定属于例外：

a) 工程师代表对任何工程或材料的不予否认，不得影响工程师此后否认以及命令拆毁、移动或拆除此种工程或材料的权力。

b) 若承包人对工程师代表的任何决定不满意，其有权将此决定提交工程师确认、取消或更改。

ASSIGNMENT AND SUB-LETTING
转让和分包

3. The Contractor shall not assign the Contract or any part thereof, or any benefit or interest therein or thereunder, otherwise than by a charge in favor of the Contractor's bankers of any monies due or to become due under this Contract, without the prior written consent of the Employer.

3. 未经业主事前书面同意，承包人不得将合同或其他任何部分，或合同所规定或依合同而产生的任何收益转让，向承包人的开户银行支付按本合同规定到期或即将到期的款项除外。

4. The Contractor shall not sub-let the whole of the Works. Except where otherwise provided by the Contract, the Contractor shall not sub-let any part of the Works without the prior written consent of the Engineer, which shall not be unreasonably withheld, and such consent, if given, shall not relieve the (contractor from any liability or obligation under the Contract and he shall be responsible for the acts, defaults and neglects of any sub-contractor, his agents, servants or workmen as fully as if they were the acts, defaults or neglects of the Contractor, his agents. servants or workmen. Provided always that the provision of labor on a piecework basis shall not be deemed to be a subletting under the Clause.

4. 承包人不得转包整个工程。除非合同另有规定，未经工程师事前书面同意，承包人也不得分包工程的任何部分，但工程师不得无故不同意分包，一旦同意分包，此种同意不得免去承包人所承担的任何合同所规定的责任或义务，他必须对任何分包人、其代理人、雇员或工人的行为、不履行和过失负完全责任，如同这些行为、不履行或过失是承包人、其代理人、雇员或工人所为。以计件方式提供劳力不得视为是本条所规定的分包。

CONTRACT DOCUMENTS
合同文件

5.1 There shall be stated in Part II of these Conditions:
   a) the language or languages in which the Contract documents shall be drawn up and
   b) the country or state, the law of which is to apply to the Contract and according to which the Contract is to be construed.

   If the said documents are written in more than one language, the language according to which the Contract is to be construed and interpreted shall also be designated in Part II, being therein designated the "Ruling Language".

5.1 以下要件得在合同第二部分规定：用以起草合同文件的语言；合同适用哪个国家的法律以及用哪个国家的法律解释合同。

   如果文件用一种以上语言制作，用以解释合同的语言也必须在第二部分中规定，且将被称为"主体语言"。

5.2 Except if and to the extent otherwise provided by the Contract, the provisions of the Conditions of Contract Parts I and II shall prevail over those of any other document forming part of the Contract. Subject to the foregoing, the several documents forming the Contract are to be taken as mutually explanatory of one another, but in case of ambiguities or discrepancies the same shall be explained and adjusted by the Engineer who shall thereupon issue to the Contractor instructions thereon. Provided always that if, in the opinion of the Engineer, compliance with any such instructions shall involve the Contractor in any cost, which by reason of any such ambiguity or discrepancy could not reasonably have been foreseen by the Contractor, the Engineer shall certify and the Employer shall pay such additional sum as may be reasonable to cover such costs.

5.2 除合同中另有规定外，合同第一、二部分的条款规定优于其他任何构成合同的文件的规定。以上述规定为准，构成合同的数个文件可视为能互相解释，如意思含糊或不一致时，由工程师解释和处理，并由此向承包人发出指令。如工程师认为，服从此种指令会使承包人发生额外费用，而此种费用是承包人由于上述意思含糊或不一致而按理无法预见的，工程师应予以证明，业主必须支付相应的额外款额以补偿此种费用。

6.1 The Drawings shall remain in the sole custody of the Engineer, but two copies thereof shall be furnished to the Contractor free of charge. The Contractor shall provide and make at his own expense any further copies required by him. At the completion of the Contract the Contractor shall return to the Engineer all Drawings provided under the Contract.

6.1 图纸由工程师独自保管，但须向承包人免费提供两份副本。承包人所需的其余副本由他自己制作并承担费用。合同履行后，承包人须将全部合同图纸归还工程师。

6.2 One copy of the Drawings, furnished to the Contractor as aforesaid, shall be kept by the contractor on the Site and the same shall at all reasonable times be available for inspection and use by the Engineer and the Engineer's Representative and by any other person authorized by the Engineer in writing.

6.2 承包人必须将按上述规定所提交的一份图纸副本留在工地，让工程师及其代表，或工程师书面授权的任何其他人在所有合理的时间内查阅使用。

6.3 The Contractor shall give written notice to the Engineer whenever planning or progress of the Works' is likely to be delayed or disrupted unless any further drawing or order, including a direction, instruction or approval, is issued by the Engineer within a reasonable time. The notice shall include details of the drawing or order required and of why and by when it is required and of any delay or disruption likely to be suffered if it is late.

6.3 如果工程师不在合理的时间内提供进一步的图纸或命令，包括指示、指令或认可，工程计划或进展便可能被延误或中断时，承包人必须书面通知工程师。通知书中应详细说明所需的图纸或命令，所需原因和时间，以及如果不及时提供而可能造成的任何延误或中断。

6.4 If, by reason of any failure or inability of the Engineer to issue within a time reasonable in all the circumstances any drawing or order requested by the Contractor in accordance with sub clause (3) of this Clause, the Contractor suffers delay and/or incurs costs then the Engineer shall take such delay into account in determining any extension of time to which the Contractor is entitled under Clause 44 hereof and the Contractor shall be paid the amount of such cost as shall be reasonable.

6.4 如承包人按本条第3款规定索要图纸或命令,由于工程师没有或不能在合理的时间内提供,从而导致承包人误工和/或增加成本,工程师必须考虑此种延误,以决定是否按本合同第44条规定延长承包人的工期,且只要有理由,承包人所承担的此种费用必须得到补偿。

7. The Engineer shall have full power and authority to supply to the Contractor from time to time, during the progress of the Works, such further drawings and instructions as shall be necessary for the purpose of the proper and adequate execution and maintenance of the Works. The Contractor shall carry out and be bound by the same.

7. 在施工期间,工程师全权负责随时向承包人提供进一步的图纸和指示,以满足工程正常施工和维护所需。承包人必须执行且受图纸和指示的约束。

GENERAL OBLIGATIONS
总义务

8.1 The Contractor shall, subject to the provisions of the Contract, and with due care and diligence, execute and maintain the Works and provide all labor, including the supervision thereof, materials, Constructional Plant and all other things, whether of a temporary or permanent nature, required in and for such execution and maintenance, so far as the necessity for providing the same is specified in or is reasonably to be inferred from the Contract.

8.1 承包人必须根据合同条款,对工程的施工和维护予以应有的注意,且提供此种施工和维护所必需的包括劳动管理在内的所有劳力、材料、施工成套设备及其他一切物品,不管其是临时或长期性质,只要合同明文规定需要或根据合同合理推断需要。

8.2 The Contractor shall take full responsibility for the adequacy, stability and safety of all site operations and methods of construction, provided that the Contractor shall not be responsible, except as may be expressly provided in the Contract, for the design or specification of the Permanent Works, or for the design or specification of any Temporary Works prepared by the Engineer.

8.2 承包人必须对现场操作和施工方法的恰当、稳定及安全性负全部责任。除非合同另有明文规定，承包人对工程师制订的永久性工程的设计或规格，或临建工程的设计或规格概不负责。

9. The Contractor shall, when called upon so to do, enter into and execute a Contract Agreement, to be prepared and completed at the cost of the Employer, in the form annexed with such modification as may be necessary.

9. 如经要求，承包人必须签署一份承包协议，该协议由业主制订并承担费用，协议应附带必要的修正条款。

10. If, for the due performance of the Contract, the Tender shall contain an undertaking by the Contractor to obtain, when required, a bond or guarantee of an insurance company or bank, or other approved sureties to be jointly and severally bound with the contractor to the Employer, in a sum not exceeding that stated in the Letter of Acceptance for such bond or guarantee, the said insurance company or bank or sureties and the terms of the said bond or guarantee shall be such as shall be approved by the Employer. The obtaining of such bond or guarantee or the provision of such sureties and the cost of the bond or guarantee to be so entered into shall be at the expense in all respects of the Contractor, unless the Contract otherwise provides.

10. 为正常履行合同，在标书中，承包人应承诺按要求取得保险公司或银行的保单或保函，或其他业经认可由承包人向业主负连带责任的担保，其数额不超过验收证书中规定的保单或保函额，上述保险公司、银行或担保以及上述保单或保函的条款必须经业主认可。此种保单或保函的取得或担保的提供，以及缔结保单或保函的费用应全部由承包人承担，合同中另有规定的除外。

11. The Employer shall have made available to the Contractor with the Tender documents such data on hydrological and sub-surface conditions as shall have been obtained by or on behalf of the Employer from investigations undertaken relevant to the Works and the Tender shall be deemed to have been based on such data, but the Contractor shall be responsible for his own interpretation thereof.

　　The Contractor shall also be deemed to have inspected and examined the Site and its surroundings and information available in connection therewith and to have satisfied himself, so far as is practicable, before submitting his Tender, as to the form and nature thereof, including the sub-surface conditions, the hydrological and climatic conditions, the extent and nature of work and materials necessary for the completion of the Works, the means of access to the site and the accommodation he may require and, in general, shall be deemed to have obtained all necessary information, subject as above mentioned, as to risks, contingencies and all other circumstances which may influence of affect his Tender.

11. 业主必须在招标文件中向承包人提供由业主或其代理人在进行工程考察时获得的水文及地质情况资料,标书必须视为是基于此种资料所制订的,但承包人必须对资料的理解自行负责。

承包人也必须被视为已视察了工地及周围环境,查阅了可获得的有关工地资料,且在提交标书前,对一切实际情况,从形式到性质,包括地质条件、水文和气候条件、工程范围和性质以及完成工程所必需的材料、到达工地的交通工具和所需的食宿等感到满意,总之,承包人必须被视为已得到所有必要的资料,除涉及上述情况外,还涉及风险、意外事件及其他一切可能影响其投标的情况。

12. The Contractor shall be deemed to have satisfied himself before tendering as to the correctness and sufficiency of his Tender for the Works and of the rates and prices stated in the priced Bill of Quantities and the Schedule of Rates and Prices, if any, which Tender rates and prices shall, except insofar as it is otherwise provided in the Contract, cover all his obligations under the Contract and all matters and things necessary for the proper execution and maintenance of the Works. if, however, during the execution of the Works the Contractor shall encounter physical conditions, other than climatic conditions on the Site, or artificial obstructions, which conditions or obstructions could, in his opinion, not have been reasonably foreseen by an experienced contractor, the Contractor shall forthwith give written notice thereof to the Engineer's Representative and if, in the opinion of the Engineer, such conditions or artificial obstructions could not have been reasonably foreseen by an experienced contractor, then the Engineer shall certify and the Employer shall pay the additional cost to which the Contractor shall have been put by reason of such conditions, including the proper and reasonable cost of complying with any instruction which the Engineer may issue to the Contractor in connection therewith, and of any proper and reasonable measures approved by the Engineer which the Contractor may take in the absence of specific instructions from the Engineer, as a result of such conditions or obstructions being encountered.

12. 承包人得被视为在投标前已对其工程标书,对标价的建筑工程清单、单价和价格表(如果有)上所列的单价和价格的正确性和完善性感到满意,此种投标价格必须贯穿其所有的合同义务,适用于所有为工程的正常施工和维护所必需的事物,除非本合同另有规定。然而,在施工期间,如承包人遇到除工地气候之外的其他自然情况或人为阻碍,依他所见,此种自然情况或人为阻碍是经验丰富的承包人也无法预见的,承包人必须立即书面通知工程师代表,如工程师确认此种情况或人为阻碍为经验丰富的承包人无法合理预见，工程师必须作证且业主支付承包人由于此种情况而承担的额外费用，包括因遇到此种情况或阻碍而为执行工程师可能向承包人发出的与此情况有关的任何指示而发生的正当合理的费用，以及在无工程师具体指示时，承包人可能采取业经工程师认可的恰当和合理措施而发生的正当合理费用。

13. Save insofar as it is legally or physically impossible, the Contractor shall execute and maintain the Works in strict accordance with the Contract to the satisfaction of the Engineer and shall comply with and adhere strictly to the Engineer's instructions and directions on any matter whether mentioned in the Contract or not, touching or concerning the Works. The Contractor shall take instructions and directions only from the Engineer or, subject to the limitations referred to in Clause 2 hereof, from the Engineer's Representative.

13. 除因法律或自然因素而不能之外，承包人必须严格按合同规定施工和维护工程，使工程师感到满意，且必须遵守和严格执行工程师有关任何事项的指令和指示，不管合同中是否有规定，提及或涉及工程，承包人只能从工程师处接受指令和指示，或根据本合同第2条规定，接受工程师代表的指令和指示。

14.1 Within the time stated in Part II of these Conditions, the Contractor shall, after the acceptance of his Tender, submit to the Engineer for his approval a program showing the order of procedure in which he proposes to carry out the Works. The Contractor shall, whenever required by the Engineer or Engineers'Representative, also provide in writing for his information a general description of the arrangements and methods which the Contractor proposes to adopt for the execution of the Works.

14.1 中标后，承包人得在第二部分条款规定的时间内向工程师提交一份其计划施工的程序方案，以征得工程师的认可。承包人还得随时应工程师或其代表的要求，提供一份承包人的施工计划安排和方法的说明书，以供其参考。

14.2 If at any time it should appear to the Engineer that the actual progress of the Works does not conform to the approved program referred to in sub-clause (1) of this Clause, the Contractorshall produce, at the request of the Engineer, a revised program showing the modifications to the approved program necessary to ensure completion of the Works within the time for completion asdefined in Clause 48 hereof.

14.2 不论何时，只要工程师发现工程实际进度与本条第1款规定的业经认可的方案不符，承包人必须应工程师的要求提供一份修改方案，对原有的方案做必要的修正，以保证工程能在本合同第48条规定的期限内完工。

14.3 The submission to and approval by the Engineer or Engineer's Representative of such programs of the furnishing of such particulars shall not relieve the Contractor of any of his duties or responsibilities under the Contract.

14.3 承包人不得因向工程师或工程师代表提交或经其认可此种方案或因提供此种细节，而被免除任何合同责任或义务。

15. The Contractor shall give or provide all necessary super-intendance during the execution of the Works and as long thereafter as the Engineer may consider necessary for the proper fulfilling of the Contractor's obligations under the Contract. The Contractor, or a competent and authorized agent or representative approved of in writing by the Engineer, which approval may at any time be withdrawn, is to be constantly on the Works and shall give his whole time to the super-intendance of the same. If such approval shall be withdrawn by the Engineer, the Contractor shall, as soon as is practicable, having regard to the requirement of replacing him as hereinafter mentioned, after receiving written notice of such withdrawal, remove the agent from the Works and shall not thereafter employ him again on the Works in any capacity and shall replace him by another agent approved by the Engineer. Such authorized agent or representative shall receive, on behalf of the Contractor, directions and instructions from the Engineer or, subject to the limitations of Clause 2 hereof, the Engineer's Representative.

15. 在施工期间，以及其后在工程师认为是为正常履行合同义务而必需时，承包人必须行使一切监督权。承包人，或其经工程师书面认可的全权代理人或代表（此种认可可随时撤销）应当随时在工地，一直进行监督管理。如工程师撤销对代理人的认可，在接到书面撤销通知后，承包人必须尽快根据撤换规定将其撤离工地，且今后不得再在工地上以任何身份雇佣他（她），且用另外一位经工程师认可的代理人予以替代。此种授权代理人或代表得代表承包人接受工程师的指令和指示，或按本合同第2条规定，接受工程师代表的指令和指示。

16.1 The Contractor shall provide and employ on the Site in connection with the execution and maintenance of the Works only such technical assistants as are skilled and experienced in their respective callings and such sub-agents, foremen and leading hands as are competent to give proper supervision to the work they are required to supervise, and such skilled, semi-skilled and unskilled labor as is necessary for the proper and timely execution and maintenance of the Works.

16.1 承包人必须在工地上提供和雇用与施工和工程维护有关的对本专业熟悉和经验丰富

的技术人员，能胜任规定监管工作的分代理人、工头和领班，以及为正常和及时施工和维护工程所需的熟练、半熟练和非熟练工人。

16.2　The Engineer shall be at liberty to object to and require the Contractor to remove forthwith from the Works any person employed by the Contractor in or about the execution or maintenance of the Works who, in the opinion of the Engineer, misconducts himself, or is incompetent or negligent in the proper performance of his duties, or whose employment is otherwise considered by the Engineer to be undesirable and such person shall not be again employed upon the Works without the written permission of the Engineer. Any person so removed from the Works shall be replaced as soon as possible by a competent substitute approved by the Engineer.

16.2　如工程师认为承包人雇来进行或有关施工或工程维护的任何人员行为不轨、或不能或疏于履行其职责、或认为其雇佣纯属不必要，工程师有权反对雇佣，并要求承包人立即将其从工地解雇，未经工程师的书面同意，此种人员不得再被雇用到工地。被解雇人员的职位应尽快由工程师认可的称职人选接替。

17.　The Contractor shall be responsible for the true and proper setting-out of the Works in relation to original points, lines and levels of reference given by the Engineer in writing and for the correctness, subject as above mentioned, of the position, levels, dimensions and alignment of all parts of the Works and for the provision of all necessary instruments, appliances and labor in connection therewith. If, at any time during the progress of the Works, any error shall appear or arise in the position, levels, dimensions or alignment of any part of the Works, the Contractor, on being required so to do by the Engineer or the Engineer's Representative, shall, at his own cost, rectify such error to the satisfaction of the Engineer or the Engineer's Representative, unless such error is based on incorrect data supplied in writing by the Engineer or the Engineer's Representative, in which case the expense of rectifying the same shall be borne by the Employer. The checking of any setting-out or of any line or level by the Engineer or the Engineer's Representative shall not in any way relieve the Contractor of his responsibility for the correctness thereof and the Contractor shall carefully protect and preserve all bench-marks, sight-rails, pegs and other things used in setting-out the Works.

17.　承包人负责按工程师提交的书面参考原图的有关点、线和面的规定，真实恰当地进行测量放线，如上所述，使工程的位置、水平、面积正确无误，并校准工程的各部分，且负责提供与工程相关的所有必要工具、设备和劳力。在施工中，如任何时候在工程的位置、水平、面积或工程任何部分的校正出现差错，应工程师或工程师代表的要求，承包人必须改正错误以使工程师或工程师代表满意，改正费用由承包人自己承担，此种错误是由工程师或工程师代表提供的书面资料差错而导致的除外，在这种情况下，须由业主承担改正费用。承包人不得因工程师或工程师代表对任何位置测量或任何线或任何水平面的检查而免去确保其正确无误的责任，承包人必须仔细保护和保存工程位置测定中使用过的水准点、观测杆、测标及其他物品。

18. If, at any time during the execution of the Works, the Engineer shall require the Contractor to make bore holes or to carry out exploratory excavation, such requirement shall be ordered in writing and shall be deemed to be an addition ordered under the provisions of Clause 51 hereof, unless a provisional sum in respect of such anticipated work shall have been included in the Bill of Quantities.

18. 在施工中，工程师如在任何时间要求承包人钻孔或进行挖掘勘探，此要求必须写成书面形式，且得视为是根据本合同第 51 条规定做出的附加命令，除非数量清单中已经列出有关此种预计工程的备用款。

19. The Contractor shall in connection with the Works provide and maintain at his own cost all lights, guards, fencing and watching when and where necessary or required by the Engineer or the Engineer's Representative, or by any duly constituted authority, for the protection of the Works, or for the safety and convenience of the public or others.

19. 凡有必要或应工程师或工程师代表、或任何正式成立的工程管理处的要求，承包人必须自费提供和维护与工程有关的所有灯光、警卫、栅栏和看护以保卫工程，或保障公众及其他人的安全和便利。

20.1 From the commencement of the Works until the date stated in the Certificate of completion for the whole of the Works pursuant to Clause 48 hereof the Contractor shall take full responsibility for the care thereof. Provided that if the Engineer shall issue a Certificate of Completion in respect of any part of the Permanent Works the Contractor shall cease to be liable for the care of that part of the Permanent Works from the date stated in the Certificate of Completion in respect of that part and the responsibility for the care of that part shall pass to the Employer. Provided further that the Contractor shall take full responsibility for the care of any outstanding work which he shall have undertaken to finish during the Period of Maintenance until such outstanding work is completed. In case any damage, loss or injury shall happen to the Works, or to any part thereof, from any cause whatsoever, save and except the excepted risks as defined in subclause

(2) of this Clause, while the Contractor shall be responsible for the care thereof the Contractor shall, at his own cost, repair and make good the same, so that at completion the Permanent Works shall be in good order and condition and in conformity in every respect with the requirements of the Contract and the Engineer's instructions. In the event of any such damage loss or injury happening from any of the excepted risks, the Contractor shall, if and to the extent required by the Engineer and subject always to the provisions of Clause 65 hereof, repair and make good the same as aforesaid at the cost of the Employer. The Contractor shall also be liable for any damage to the Works occasioned by him in the course of any operations carried out by him for the purpose of completing any outstanding work or complying with his obligations under Clauses 49 or 50 hereof.

20.1 从开工至按本合同第 48 条中的竣工证书规定的日期为止，承包人都得对工程全权负责。只要工程师就永久性工程的任何部分签发了竣工证书，承包人从部分竣工证书中规定的日期起不再对永久性工程的此部分负责，此部分的责任则转至业主。此外，承包人必须对任何尚未完工而他得在维护期内完成的工程进行全权维护，直至此工程完工。如工程或其任何部分出现任何损害、损失或毁坏，不论何种原因，除本条第 2 款规定的除外风险外，承包人均得自费负责修理和修补，以确保永久性工程竣工时处于状态良好，各方面都合乎合同的要求和工程师的指示。如因除外风险而导致任何损害、损失或毁坏发生，承包人必须应工程师的要求（如果有）以及本合同第 65 条的规定，如上所述，进行修理和修补，费用由业主承担。承包人也必须对为完成未竣工程或履行本合同第 49 或 50 条规定的义务，在施工过程中对工程造成的任何损害负责。

20.2 "The excepted risks" are war, hostilities (whether war be declared or not), invasion act of foreign enemies, rebellion, revolution, insurrection or military or usurped power, civil war, or unless solely restricted to employees of the Contractor or of his sub-contractors and arising from the conduct of the Works, riot, commotion or disorder, or use or occupation by the Employer of any part of the Permanent Works, or a cause solely due to the Engineer's design of the Works, or ionizing radiation, or contamination by radio-activity from any nuclear fuel or from any nuclear waste from the combustion of nuclear fuel, radio-active toxic explosive, or other hazardous properties of any explosive, nuclear assembly or nuclear component thereof, pressure waves caused by aircraft or other aerial devices travelling at sonic or supersonic speeds, or any such operation of the forces of nature as an experienced contractor could not foresee, or reasonably -make provision for or insure against all of which are herein collectively referred to as "the excepted risks".

20.2 "除外风险"包括战争,敌对状态(无论是否宣战),侵略,外国敌人行为,叛乱,革命,起义或兵变或篡权,内战,或不是由承包商的雇员、其转包人单独制造和不是因工程管理而发生的暴动、骚乱或混乱,或因业主使用或占用任何部分的永久性工程,或纯属工程师工程设计的原因,或因任何核燃料或核燃料燃烧后的废料以及放射性有毒爆炸物引起的辐射和污染,或任何爆炸物,核装置或核装置部件的其他危险特性,以音速、超音速飞行的飞机或其他飞行物的压力波以及其他任何此种自然力的作用,其不能为有经验的承包商所预见,也不能合理提供物资或投保与之对抗,所有这一切在本合同中都被称为"除外风险"。

21. Without limiting his obligations and responsibilities under Clause 20 hereof, the Contractor shall insure in the joint names of the Employer and the Contractor against all loss or damage from whatever cause arising, other than the excepted risks, for which he is responsible under the terms of the Contract and in such manner that the Employer and Contractor are covered for the period stipulated in Clause 20 (1) hereof and are also covered during the Period of Maintenance for loss or damage arising from a cause, occurring prior to the commencement of the Period of Maintenance, and for any loss or damage occasioned by the Contractor in the course of any operations carried out by him for the purpose of complying with his obligations under Clauses 49 and 50 hereof:

   a) The Works for the time being executed to the estimated current contract value thereof, or such additional sum as may be specified in Part Ⅱ in the Clause numbered 21, together with the materials for incorporation in the works at their replacement value.

   b) The Constructional Plant and other things brought on to thethe replacement value of such Constructional Plant and other things. Such insurance shall be effected with an insurer and in terms apwhich approval shall not be unreasonably withheld, and the Contractor shall, whenever required, produce to the Engineer or the Engineer's Representative the policy or policies of insurance and the receipts for payment of the current premiums.

21. 承包人必须以业主和承包人的共同名义,为防止合同规定的应由承包人负责的、除外风险外的一切损失或损害投保,不论其由什么原因造成,此种投保不得减少本合同第20条所规定的承包人的义务和责任,业主和承包人的保险除包括本合同第20.1条规定的期限外,还包括因维护期开始前发生的原因而在维护期内产生的损失或损害,以及承包人在履行本合同第49或50条规定的义务而开展任何工作期间所造成的任何损失或损害。投保项目包括:

a) 正在施工的工程，按其目前的合同价值，或按 21 条第 2 款可能规定的附加款额，连同按替换价值计算的用于工程的材料。

b) 承包商带到工地上的建筑成套设备和其他物品，按其替换价值投保。此种保险必须在一保险公司投放，条款得经业主认可，业主不得无故不同意投保，承包人必须随时应要求向工程师或工程师代表出示保险单和支付现行保险费的收据。

22.1 The Contractor shall, except if and so far as the Contract provides otherwise, indemnify the Employer against all losses and claims in respect of injuries or damage to any person or material or physical damage to any property whatsoever which may arise out of or in consequence of the execution and maintenance of the Works and against all claims, proceedings, damages, costs, charges and expenses whatsoever in respect of or in relation thereto except any compensation or damages for or with respect to:

a) The permanent use or occupation of land by the Works or any part thereof.

b) The right of the Employer to execute the Works or any part thereof on, over, under in or through any land.

c) Injuries or damage to persons or property which are the unavoidable result of the execution or maintenance of the Works in accordance with the Contract.

d) Injuries or damage to persons or property resulting from any act or neglect of the Employer, his agents, servants or other contractors, not being employed by the Contractor, or for or in respect of any claims, proceedings, damages, costs, charges and expenses in respect thereof or in relation thereto or where the injury or damage was contributed to by the Contractor, his servants or agents such part of the compensation as may be just and equitable having regard to the extent of the responsibility of the Employer, his servants or agents or other contractors for the damage or injury.

22.1 除合同另有规定外，承包人必须保护业主不得因施工和工程维护而产生或导致的任何人员伤害、材料损失及财产损失而受任何损失和做任何赔偿，且不因所有与之有关的任何索赔、诉讼、损害赔偿金、诉讼费、开支和费用而受损失，对下列事项所作或与之有关的补偿或损害赔偿除外：

a) 工程或部分工程永久使用或占用土地。

b) 业主在任何土地面上、上方、下面、里面或经过部分施工或部分施工的权利。

c) 按合同规定施工或维护工程而不可避免的人身伤害或财产损失。

d) 因业主、其代理人、雇员或其他不为承包人所雇用的承包商的任何行为或过失所造成的人员伤害或财产损失，或与之有关的任何索赔、诉讼、损害赔偿金、诉讼费、开支和费用，或曾由承包人、其雇员或代理人承担的，但原本应当由业主、其雇员或代理人或其他承包商负责的那部分涉及损失或伤害的赔偿。

22.2　The Employer shall indemnify the Contractor against all claims, proceedings, damages, costs, charges and expenses in respect of the matters referred to in the provision sub-clause (1) of this Clause.

22.2　业主必须保护承包人不因与本条第1款规定事项有关的一切索赔、诉讼、损害赔偿金、诉讼费、开支和费用而受损失。

23.1　Before commencing the execution of the Works the Contractor, but without limiting his obligations and responsibilities under Clause 22 hereof, shall insure against his liability for any material or physical damage, loss or injury which may occur to any property, including that of the Employer, or to any person, including any employee of the Employer, by or arising out of the execution of the Works or in the carrying out of the Contract, otherwise than due to the matters referred to in the provision Clause 22 (1) hereof.

23.1　在开工前，在不减少本合同第22条规定给他的义务和责任条件下，承包人必须对任何可能由于或因施工或因履行本合同而给包括业主的财产在内的任何财产，以及给包括业主的雇员在内的任何人员造成的重大或实质性损害、损失或伤害进行责任保险，本合同第22条第1款规定的事项除外。

23.2　Such insurance shall be effected with an insurer and in terms approved by the Employer, which approval shall not be unreasonably withheld, and for at least the amount stated in the Appendix to the Tender. The Contractor shall, whenever required, produce to the Engineer or the Engineer's Representative the policy or policies of insurance and the receipts for payment of the current premiums.

23.2　此种保险必须在一保险公司投放，条款得经业主认可，业主不得无故不同意投保，保险金额不得少于标书附件规定的数额。承包人必须随时应要求向工程师或工程师代表出示保险单和支付现行保险费的收据。

23.3　The terms shall include a provision whereby, in the event of any claim in respect of which the Contractor would be entitled to receive indemnity under the policy being brought or made against the Employer, the insurer will indemnify the employer against such claims and any costs, charges and expenses in respect thereof.

23.3　保险条款中必须规定，承包人不得对可能得到保险赔偿的有关事项对业主提出任何索赔，保险公司应保护业主免受索赔损失，且赔偿其有关的任何诉讼费及开支和费用。

24.1　The Employer shall not be liable for or in respect of any damages or compensation payable at law in respect or in consequence of any accident or injury to any workman or other person in the employment of the Contractor or any sub-contractor, save and except an accident or injury resulting from any act or default of the Employer, his agents, or servants. The Contractor shall indemnify and keep indemnified the Employer against all such damages and compensation, save and except as aforesaid, and against all claims, proceedings, costs, charges and expenses whatsoever in respect thereof or in relation thereto.

24.1 业主不对因承包人或任何转包人的工人或其他雇用人员的任何事故或伤害而根据法律应予支付的任何赔偿金负责,除非事故或伤害是由业主、其代理人或雇员的任何行为或玩忽职守引起的。除上述规定外,承包人得保护业主不因所有此种损害赔偿,以及因与此有关的所有索赔、诉讼、诉讼费、开支和费用而受损失。

24.2 The Contractor shall insure against such liability with an insurer approved by the Employer, which approval shall not be unreasonably withheld, and shall continue such insurance during the whole of the time that any persons are employed by him on the Works and shall, when required, produce to the Engineer or the Engineer's Representative such policy of insurance and the receipt for payment of the current premium. Provided always that, in respect of any persons employed by any sub-contractor, the Contractor's obligation to insure as aforesaid under this sub-clause shall be satisfied if the sub-contractor shall have insured against the liability in respect of such persons in such manner that the Employer is indemnified under the policy, but the Contractor shall require such sub-contractor to produce to the Engineer or the Engineer's Representative, when required, such policy of insurance and the receipt for the payment of the current premium.

24.2 承包人必须就此种责任投保,保险公司得经业主同意,业主不得无故不予同意,承包人必须在工地雇用工人的整个期间继续保险,并随时应要求向工程师或工程师代表出示此种保险单和支付现行保险费的收据。就任何转包人所雇用的人员,如转包人已经就此种人员的责任投保,且以业主为保险补偿对象,则承包人上述的投保义务被视作已经履行,但承包人必须要求此转包人随时应要求向工程师或工程师代表出示此种保险单和支付现行保险费的收据。

25. If the Contractor shall fail to effect and keep in force the insurance referred to in Clauses 21, 23 and 24 hereof, or any other insurance which he may be required to effect under the terms of the Contract, then and in any such case the Employer may effect and keep in force any such insurance and pay such premium or premiums as may be necessary for that purpose and from time to time deduct the amount so paid by the Employer as aforesaid from any monies due or which may become due to the Contractor, or recover the same as a debt due from the Contractor.

25. 如果承包人未按本合同第 21、23 和 24 条的规定投保和继续保险，或未按本合同条款规定就应由他投的其他任何保险进行投保，在任何此种情况下，业主可购买和继续任何保险，支付此种必要的费用，且随时从应付或可能应付给承包人的款项中将业主上述所支付的费用扣除，或将此作为承包人的负债追偿。

26.1 The Contractor shall give all notices and pay all fees required to be given or paid by any National Or State Statute, Ordinance, or other Law, or any regulation, or by-law of any local or other duly constituted authority in relation to the execution of the Works and by the rules and regulations of all public bodies and companies whose property or rights are affected or may be affected in any way by the Works.

26.1 承包人必须按国家或州的一切有关施工的法令、条例、其他法规、任何条例、当地或其他合法当局的地方法规以及所有其财产或权利受到或可能受到工程影响的公共团体和公司的规章制度的规定，发布所有的通告和支付所有的费用。

26.2 The Contractor shall conform in all respects with the provisions of any such Statute, Ordinance or Law as aforesaid and the regulations or by-laws of any local or other duly constituted authority which may be applicable to the Works and with such rules and regulations of public bodies and companies as aforesaid and shall keep the Employer indemnified against all penalties and liability of every kind for breach of any such Statute, Ordinance or Law, regulation or by-law.

26.2 承包人必须在所有方面遵守以上任何法令、条例或法规，可适用于工程的当地或其他合法当局的一切地方法规，以及上述公共团体和公司的规章制度，如违反上述任何法令、条例、法规或地方法规，不得让业主受任何惩罚和承担任何责任。

26.3 The Employer will repay or allow to the Contractor all such sums as the Engineer shall certify to have been properly payable and paid by the Contractor in respect of such fees.

26.3 如工程师证实业主应偿还已由承包人支付的此种费用，业主应偿付或同意偿付承包人所有此种金额。

27. All fossils, coins, articles of value or antiquity and structures and other remains or things of geological or archaeological interest discovered. the Works shall as between the Employer and the Contractor be deemed to be the absolute property of the Employer. The Contractor shall take reasonable precautions to prevent his workmen or any other persons from removing or damaging any such article or thing and shall immediately upon discovery thereof and, before removal, acquaint the Engineer's Representative of such discovery and carry out, at the expense of the Employer, the

Engineer's Representative's orders as to the disposal of the same.

27. 所有在工地发现的化石、钱币、有价值的物品或古物、建筑物和其他具有地质或考古价值的遗址或物品，均应被业主和承包人共同认定为是业主的绝对财产。承包人必须采取恰当的措施，防止其工人或其他任何人转移或破坏任何此种物品，并在物品发现后不得转移，立即通知工程师代表并执行工程师代表的有关处置命令，其费用由业主承担。

28. The Contractor shall save harmless and indemnify the Employer from and against all claims and proceedings for or on account of infringement of any patent rights, design, trademark or name or other protected rights in respect of any Constructional Plant, machine work, or material used for or in connection with the Works or any of them and from and against all claims, proceedings, damages, costs, charges and expenses whatsoever in respect thereof or in relation thereto. Except where otherwise specified, the Contractor shall pay all tonnage and other royalties, rent and other payments or compensation, if any, for getting stone, sand, gravel clay or other materials required for the Works or any of them.

28. 承包人必须保护业主不得因在工程或部分工程中使用或涉及任何建筑设备、机械加工或材料，从而侵犯与之有关的专利权、外观设计、商标或商号或其他应受保障的权利而遭索赔或被起诉，并必须保护业主免受与此有关的任何索赔、诉讼、损害赔偿金、诉讼费、支出或费用的损害。除另有规定外，承包人必须承担工程或部分工程所需的石头、沙子、砂砾、泥土或其他材料的运费、沙石产地使用费、租金和其他费用或补偿（如果有）。

29. All operations necessary for the execution of the Works shall, so far as compliance with the requirements of the Contract permits, be carried on so as not to interfere unnecessarily or improperly with the convenience of the public, or the access to, use and occupation of public or private roads and footpaths to or of properties whether in the possession of the Employer or of any other person. The Contractor shall save harmless and indemnify the Employer in respect of all claims, proceedings, damages, costs, charges and expenses whatsoever arising out of, or in relation to, any such matters in so far as the Contractor is responsible therefor.

29. 所有施工必需的活动，只要符合合同许可证的规定，都必须开展，保证不无故妨碍公众的便利，或妨碍公共或私人道路的通行、使用和占用，无论道路是通向或是在业主或其他人的地产上。倘若发生此类应由承包人负责的事件，承包人必须保护业主免受由此而引而起的任何索赔、诉讼、损害赔偿费、诉讼费、开支和费用的损害。

30. 1 The Contractor shall use every reasonable means to prevent any of the highways or bridges communicating with or on the routes to the Site from being damaged or

injured by any traffic of the Contractor or any of his sub-contractors and, in particular, shall select routes, choose and use vehicles and restrict and distribute loads so that any such extraordinary traffic as will inevitably arise from the moving of plant and material from and to the Site shall be limited as far as reasonably possible, and so that no unnecessary damage or injury may be occasioned to such highways and bridges.

30.1 承包人必须采取各种合理手段防止连接或通向工地的任何公路或桥梁由于承包人或其任何转包人的使用而遭受损害，承包人尤其必须挑选路线，选择使用车辆，限制和分散装载量，从而尽可能合理限制因往来于工地的设备和材料的运输而必然造成的异常交通，由此避免对公路和桥梁造成不必要的损害。

30.2 Should it be found necessary for the Contractor to move one or more loads of Constructional Plant, machinery or pre-constructed units or parts of units of work over part of a highway or bridge, the moving whereof is likely to damage any highway or bridge unless special protection or strengthening is carried out, then the Contractor shall before moving the load on to such highway or bridge give notice to the Engineer or Engineer's Representative of the weight and other particulars of the load to be moved and his proposals for protecting or strengthening the said highway or bridge. Unless within fourteen days of the receipt of such notice the Engineer shall by counter notice direct that such protection or strengthening is unnecessary, then the Contractor will carry out such proposals or any modification thereof that the Engineer shall require and, unless there is an item or are items in the Bill of Quantities for pricing by the Contractor of the necessary works for the protection or strengthening aforesaid, the costs thereof shall be paid by the Employer to the Contractor.

30.2 如承包人必须经某段公路或桥梁一次或多次运送建筑设备、机械或工程预制件或预制部件，如不采取特殊保护和加固措施，此种运送很可能损坏公路或桥梁，承包人运送货物通过此公路或桥梁前必须通知工程师或工程师代表，告之运送货物的重量及其他具体情况，以及其拟对上述公路或桥梁采取的保护和加固措施。除非在接到通知14天内，工程师以取消通知的方式指示不必采取此种保护或加固措施，承包人将施行此种方案或施行经工程师要求修改后的方案，除非建筑工程清单中含有一项或多项已由承包人标价的上述公路或桥梁保护或加固的必要项目，其费用得由业主向承包人支付。

30.3 If during the execution of the Works or at any time thereafter the contractor shall receive any claim arising out of the execution of the Works in respect of damage or injury to highways or bridges he shall immediately report the sums to the Engineer and thereafter the Employer shall negotiate the settlement of and pay all sums due in respect of such claim and shall indemnify the Contractor in respect thereof and in respect of all claims, proceedings, damages, Costs, charges and expenses in relation thereto. Provided always that if and so far as any such claims or part thereof shall in the opinion of the Engineer be due to any failure on the part of the Contractor to observe and perform his obligations under sub-clauses (1) and (2) of this Clause, then the amount certified by the Engineer to be due to such failure shall be paid by the Contractor to the Employer.

30.3　如果在施工期间或以后任何时间，承包人因施工损坏公路或桥梁而被索赔，他必须立即报告工程师，其后，业主将商洽解决并支付索赔款，并保护承包人免受该索赔以及与之有关的一切索赔、诉讼、损害赔偿金、诉讼费、开支和费用的损害。如果工程师认为造成此种索赔或部分索赔的原因在于承包人没有履行本条第 1 款和第 2 款规定的义务，由工程师确认的因此种不履行而损失的款额必须由承包人支付给业主。

30.4　Where the nature of the Works is such as to require the use by the Contractor of water borne transport the foregoing provisions of this Clause shall be construed as though "highway" included a lock, dock, sea wall or other structure related to a waterway and "vehicle" included craft, and shall have effect accordingly.

30.4　如果工程性质需要承包人使用水路运输，本条上述规定必须按"公路"包括船闸、码头、防波堤或其他有关的水路设施，"运输工具"包括船只来予以解释，并由此而施行。

31.　The Contractor shall, in accordance with the requirements of the Engineer, afford all reasonable opportunities for carrying out their work to any other contractors employed by the Employer and their workmen and to the workmen of the Employer and of any other duly constituted authorities who may be employed in the execution on or near the Site of any work not included in the Contract or of any contract which the Employer may enter into in connection with or auxiliary to the Works. if, however, the Contractor shall, on the written request of the Engineer or the Engineer's Representative, make available to any such other contractor, or to the Employer or any such authority, any roads or ways for the maintenance of which the Contractor is responsible, or permit the use by any such of the contractor's scaffolding or other plant on the Site, or provide any other service of whatsoever nature for any such, the Employer shall pay to the Contractor in respect of such use or service such sum or sums as shall, in the opinion of the Engineer, be reasonable.

31. 承包人必须按工程师的要求为业主雇用的其他承包商及其工人,以及为业主或其他合法当局雇用的工人提供合理开展工作的机会,这些工人可受雇在工地上或附近,从事不为本合同规定的任何工作或由业主签订的与工程有关或附属工程的任何其他合同规定的工作。然而,如承包人应工程师或工程师代表的书面要求,为其他承包人、业主或其他合法当局提供由他负责维护的道路,或准许使用其在工地的脚手架或其他设备,或提供其他任何类似性质的服务,只要工程师认为合理,业主必须向承包人支付有关此种使用或服务的费用。

32. During the progress of the works the Contractor shall keep the Site reasonably free from all unnecessary obstruction and shall store or dispose of any Constructional Plant and surplus materials and clear away and remove from the Site any wreckage, rubbish or Temporary Works no longer required.

32. 在施工期间,承包人应保证工地没有任何不必要的障碍物,妥善储存或处置建筑设备和多余的材料,将所有残余物垃圾或不再需要的临时工程设施清理出工地。

33. On the completion of the Works the Contractor shall clear away and remove from the Site all Constructional Plant, surplus materials, rubbish and Temporary Works of every kind, and leave the whole of the Site and Works clean and in a workmanlike condition to the satisfaction of the Engineer.

33. 完工之后,承包人应将所有建筑设备、多余材料、垃圾和各种临时工程设施清理出工地,使整个工地和工程显得整洁,合乎标准,让工程师满意。

LABOR
劳工

34.1 The Contractor shall make his own arrangements for the engagement of all labor, local or otherwise, and save insofar as the Contract otherwise provides, for the transport, housing feeding and payment thereof.

34.1 承包人必须自己安排雇用所有当地或其他地方的劳工,并负责劳工的交通、食宿和工资,本合同另有规定的除外。

34.2 The Contractor shall, so far as is reasonably practicable, having regard to local conditions, provide on the Site, to the satisfaction of the Engineer's Representative, an adequate supply of drinking and other water for the use of the Contractor's staff and work people.

34.2 只要合理可行,承包人应视当地情况向工地的职员和工人提供充足的饮水和其他用水,让工程师代表感到满意。

34.3 The Contractor shall not, otherwise than in accordance with the Statutes, Ordinances and Government Regulations or Orders for the time being in force, import, sell, give, barter or otherwise dispose of any alcoholic liquor, or drugs, or permit or suffer any such importation, sale, gift, barter or disposal by his sub-contractors, agents or employees.

34.3 除非根据现行法律、法规及政府规章或命令，承包人不得进口、销售、赠予、以物交换或以其他方式处置任何烈酒或毒品，或允许、容忍转包人、代理人或雇员参与任何此种物品的进口、销售、赠予、交换或处置。

34.4 The Contractor shall not give, barter or otherwise dispose of to any person or persons, any arms or ammunition of any kind or permit or suffer the same as aforesaid.

34.4 如上所述，承包人不得向他人赠予、交换或以其他方式处置任何武器或弹药，也不得准许或容忍同样行为发生。

34.5 The Contractor shall in all dealings with labor in his employment have due regard to all recognized festivals, days of rest and religious or other customs.

34.5 在处理与雇工的关系时，承包人应适当注意公认的节假日和宗教或其他习俗。

34.6 In the event of any outbreak of illness of an epidemic nature, the Contractor shall comply with and carry out such regulations, orders and requirements as may be made by the Government, or the local medical or sanitary authorities for the purpose of dealing with and overcoming the same.

34.6 一旦爆发传染病，承包人必须遵守并执行政府或当地医疗卫生机构为治疗此种疾病所制定的规章、命令及要求。

34.7 The Contractor shall at all times take all reasonable precautions to prevent any unlawful, riotous or disorderly conduct by or amongst his employees and for the preservation of peace and protection of persons and property in the neighborhood of the Works against the same.

34.7 承包人应随时采取一切合理的预防措施，防止其雇员制造或在雇员中发生违法、骚乱或混乱行为，以维护治安和保护工程附近地区人员和财产的安全。

34.8 The Contractor shall be responsible for observance by his sub-contractors of the foregoing provisions.

34.8 承包人必须负责让其转包人遵守上述规定。

34.9 Any other conditions affecting labor and wages shall be as set out in part Ⅱ in the clause numbered 34 as may be necessary.

34.9 如有必要，可在第二部分的第34条就所有其他有关劳工和工资的条件做出规定。

35. The Contractor shall, if required by the Engineer, deliver to the Engineer Representative, or at his office, a return in detail in such form and at such intervals as the Engineer may prescribe showing the supervisory staff and the numbers of the several classes of labor from time to time employed by the Contractor on the Site and such information respecting Constructional Plant as the Engineer's Representative may require.

35. 如工程师要求，承包人应按工程师要求的格式和时间向工程师代表或其办公室提交一份详细报告，汇报其在工地随时雇用的监督人员和不同等级劳工的数量，以及工程师代表可能要求的有关建设设备的情况。

MATERIALS AND WORSHIP
材料和工艺

36.1 All materials and workmanship shall be of the respective kinds described in the Contract and in accordance with the Engineer's instructions and shall be subjected from time to time to such tests as the Engineer may direct at the place of manufacture or fabrication, or on the Site or at such other place or places as may be specified in the Contract, or at all or any of such places. The Contractor shall provide such assistance, instruments, machines, labor and materials as are normally required for examining, measuring and testing any work and the quality, weight or quantity of any material used and shall supply samples of materials before incorporation in the Works for testing as may be selected and required by the Engineer.

36.1 所有材料和工艺均应符合合同的规定和工程师的指令，且应随时在制造或加工地、工地或合同具体规定的其他地点，同时或在某一处，接受工程师可能命令进行的检验。承包人要为任何工程或任何材料的质量、重量或数量的检查、测量和检验提供正常所需的帮助、工具、机械、劳力和材料，承包人还必须应工程师可能的选择和要求，在材料应用于工程之前提供材料样品以供检验。

36.2 All samples shall be supplied by the Contractor at his own cost if the supply thereof is clearly intended by or provided for in the Contract, but if not, then at the cost of the Employer.

36.2 如果合同明显默示或明文规定，所有由承包人提供的样品的费用应由承包人自己承担，否则应由业主支付。

36.3 The cost of making any test shall be borne by the Contractor if such test is clearly intended by or provided for in the Contract and, in the cases only of a test under load or of a test to ascertain whether the design of any finished or partially finished work is appropriate for the purposes which it was intended to fulfil, is particularized in the Contract in sufficient detail to enable the contractor to price or allow for the same in his Tender.

36.3 如果合同明显默示或明文规定，一切检验费用应由承包人负担，对于仅是负载检测和确认任何完成或部分完成的工程的设计是否符合要求的检验，本合同做有详尽规定，可确保或准许承包商将其费用计入投标价格中。

36.4 If any test is ordered by the Engineer which is either not、intended by or provided for, or (in the cases above mentioned) is not so particularized, or though so intended or provided for is ordered by the Engineer to be carried out by an independent person at any place other than the Site or the place of manufacture or fabrication of the materials tested, then the cost of such test shall be borne by the Contractor, if the test shows the workmanship or materials not to be in accordance with the provisions of the Contract or the Engineer's instructions, but otherwise by the Employer.

36.4 凡是工程师命令所做的检验，而合同没有默示或规定，或（在上述情况中）没有如此详细规定，或虽有默示或规定，但却是由工程师命令独立个人在工地之外，或在该检验物资的制造或加工地之外的任何地点进行的，如检验结果表明工艺或材料不符合合同规定或工程师指令，检验费由承包人承担，否则由业主支付。

37. The Engineer and any person authorized by him shall at all times have access to the Works and to all workshops and places where work is being prepared or from where materials, manufactured articles or machinery are being obtained for the Works and the Contractor shall afford every facility for and every assistance in or in obtaining the right to such access.

37. 工程师和他授权的任何人随时有权到工程现场，到任何工程准备地或工程所需材料、制品或机械获得地，承包人应为其行使或获得此种到场权利提供一切便利和帮助。

38.1 No work shall be covered up or put out of view without the approval of the Engineer or the Engineer's Representative and the Contractor shall afford full opportunity for the Engineer or the Engineer's Representative to examine and measure any work which is about to be covered up or put out of view and to examine foundations before permanent work is placed thereon. The Contractor shall give due notice to the Engineer's Representative whenever any such work or foundations is or are ready or about to be ready for examination and the Engineer's Representative shall, without unreasonable delay, unless he considers it unnecessary and advises the Contractor accordingly, attend for the purpose of examining and measuring such work or of examining such foundations.

38.1 未经工程师或工程师代表的认可，任何工程不得被覆盖或遮掩，承包人应为工程师或工程师代表提供充分机会，以检查和测量即将被覆盖或遮掩的任何工程，以及在永久性工程开始前检查地基。承包人应在任何此种工程或地基做好或即将做好检查准备时及时通知工程师代表，工程师代表必须参加此种工程或地基的检查，不得无故拖延，除非他认为检查没有必要，且由此通知承包人。

38.2 The Contractor shall uncover any part or parts of the works or make openings in or through the same as the Engineer may from time to time direct and shall reinstate and make good such part or parts to the satisfaction of the Engineer. If any such part or parts have been covered up or put out of view after compliance with the requirement of sub-clause (1) of this Clause and are found to be executed in accordance with the Contract, the expenses of uncovering, making openings in or through, reinstating and making good the same shall be borne by the Employer, but in any other case all costs shall be borne by the Contractor.

38.2 承包人必须应工程师随时做出的指示，打开已覆盖的工程的某一部分或某些部分，或在工程上或经过工程开通道，并负责使之完好复原而让工程师满意。凡任何此种工程部分是在履行本条第 1 款规定后被覆盖或遮掩的，则打开和在工程上或经过工程开通道，以及使之完好复原的费用由业主承担，否则所有费用应由承包人承担。

39.1 The Engineer shall during the progress of the Works have power to order in writing from time to time: the removal from the Site, within such time or times as may be specified in the order, of any materials which, in the opinion of the Engineer, are not in accordance with the Contract, the substitution of proper and suitable materials and, the removal and proper re-execution, notwithstanding any previous test thereof or interim payment therefor, of any work which in respect of materials or workmanship is not, in the opinion of the Engineer, in accordance with the Contract.

39.1 在施工中，工程师有权随时书面命令：把工程师认为不符合合同规定的任何材料按命令中具体规定的时间从工地换走，用恰当合适的材料予以替换和把工程师认为不符合合同规定的材料或工艺换掉或适当重新施工，无论其在之前是否经过检查或做过中期支付。

39.2 In case of default on the part of the Contractor in carrying out such order, the Employer shall be entitled to employ and pay other persons to carry out the same and all expenses consequent hereon or incidental thereto shall be recoverable from the Contractor by the Employer, or may be deducted by the Employer from any monies due or which may become due to the Contractor.

39.2 如承包人未执行此种命令，业主有权雇用他人执行，业主应从承包人处收回由此而产生的所有费用和杂费，或从应付给或即将付给承包商的款项中予以扣除。

40.1 The Contractor shall, on the written order of the Engineer, suspend the progress of the Works or any part thereof for such time or times and in such manner as the Engineer may consider necessary and shall during such suspension properly protect and secure the work, so far as is necessary in the opinion of the Engineer. The extra cost incurred by the Contractor in giving effect to the Engineer's instructions under this Clause shall be borne and paid by the Employer unless such suspension is: a) otherwise provided for in the Contract, or b) necessary by reason of some default on the part of the Contractor, or c) necessary by reason of climatic conditions on the Site, or d) necessary for the proper execution of the Works or for the safety of the Works or any part thereof insofar as such necessity does not arise from any act or default by the Engineer or the Employer or from any of the excepted risks defined in Clause 20 hereof.

Provided that the Contractor shall not be entitled to recover any such extra cost unless he gives written notice of his intention to claim to the Engineer within _____ days of the Engineer's order. The Engineer shall settle and determine such extra payment and/or extension of time under Clause 44 hereof to be made to the Contractor in respect of such claim as shall, in the opinion of the Engineer, be fair and reasonable.

40.1 收到工程师的书面命令后，承包人必须在工程师认为必要的时间以其认为必要的方式停止工程或部分工程的施工，只要工程师认为必要，还必须在停工期间妥善保护工程的安全。承包人根据本条款执行工程师指令而发生的额外费用应由业主承担，除非此种停工是合同另行规定的，或由承包人的某种违约而导致的必要停工，或由工地的气候条件导致的必要停工，或为正常施工或为整个工程或部分工程的安全而做的必要停工，而不是因工程师或业主的任何行为或违约而产生的，也不是因本合同第20条规定的任何除外风险引起的。

承包人在收到工程师命令后_____天内应书面通知工程师要求赔偿，否则将无权得到额外费用。如工程师认为承包人的要求公平合理，他必须按本合同第44条的规定，处理并决定此种支付给承包人的额外费用和/或停工时间。

40.2 If the progress of the Works or any part thereof is suspended on the written order of the Engineer and if permission to resume work is not given by the Engineer within a period of ninety days from the date of suspension then, unless such suspension is within paragraph a), b), c) or d) of sub—clause (1) of this Clause, the Contractor may serve a written notice on the Engineer requiring permission within _____ days from the receipt thereof to proceed with the Works, or that part thereof in regard to which progress is suspended and, if such permission is not granted within that time, the Contractor by a further written notice so served may, but is not bound to, elect or treat the suspension where it affects part only of the works as an omission of such part under Clause 51 hereof, or, where it affects the whole Works, as an abandonment of the Contract by the Employer.

40.2 如整个工程或部分工程按工程师的命令停工，且停工 90 天后仍未得到工程师做出的复工命令，除非此种停工属本条第 1 款第 4 项规定范围之内，否则承包人可书面通知工程师，要求其在收到通知后 _____ 天内准许对整个工程或停工的部分工程重新开工，如在此期间仍未获得开工许可，承包人可以，但不一定必须再做书面通知，按本合同第 51 条的规定将只影响到部分工程的停工视为是对该部分工程的省略，或如停工影响到整个工程，视其为业主废弃合同。

COMMENCEMENT TIME AND DELAYS
开工时间和延误

41. The Contractor shall commence the Works on Site within the period named in the Appendix to the Tender after the receipt by him of a written order to this effect from the Engineer and shall proceed with the works with due expedition and without delay, except as may be expressly sanctioned or ordered by the Engineer, or be wholly beyond the Contractor's control.

41. 接到工程师的书面开工通知后，承包人应在标书附录中规定的期限内在工地施工，且必须迅速，不得延误，工程师应明文同意或命令，或承包人无法控制的原因除外。

42.1 Save insofar as the Contract may prescribe, the extent of portions of the Site of which the Contractor is to be given possession from time to time and the order in which such portions shall be made available to him and, subject to any requirement in the Contract as to the order in which the works shall be executed, the Employer will, with the Engineer's written order to commence the Works, give to the Contractor possession of so much of the site as may be-required to enable the Contractor to commence and proceed with the execution of the Works in accordance with the program referred to in Clause 14 hereof, if any, and otherwise in accordance with such reasonable proposals of the Contractor as he shall, by written notice to the Engineer, make and will, from time to time as the Works proceed, give to the Contractor possession of such further portions of the Site as may be required to enable the Contractor to proceed with the execution of the works with due dispatch in accordance with the said program or proposals, as the case may be. If the Contractor suffers delay or incurs cost from failure on the part of the Employer to give possession in accordance with the terms of this Clause, the Engineer shall grant an extension of time for the completion of the Works and certify such sum as, in his opinion, shall be fair to cover the cost incurred, which sum shall be paid by the Employer.

42.1 除合同另有规定的外，就承包人应随时获得的工地的范围及授权他使用此部分工地的命令而言，业主应根据合同有关施工命令的规定，随同工程师的书面开工命令，提供承包人必要范围的工地，使其能够开始并按本合同第 14 条规定的方案施工，此外应根据承包人给工程师的书面通知上的合理建议（如果有），在施工期间随时向承包人提供更大范围的工地，以保证承包人能按上述方案或建议迅速施工。如承包人因业主不按本条规定提供工地而误工或发生费用，工程师应准许延长工期，并由他确定一笔补偿此种费用的款项，该款项应由业主支付。

42.2 The Contractor shall bear all costs and charges for special or temporary way leaves required by him in connection with access to the site. The Contractor shall also provide at his own cost any additional accommodation outside the Site required by him for the purposes of the Works.

42.2 承包人应承担因通往工地的所有特殊或临时道路通行权而产生的费用。承包人也得支付为施工而应他的要求在工地之外食宿而发生的一切额外费用。

43. Subject to any requirement in the Contract as to completion of any section of the Works before completion of the whole, the whole of the Works shall be completed, in accordance with the provisions of Clause 48 hereof, within the time stated in the Contract calculated from the last day of the period named in the Appendix to the Tender as that within which the Works are to be commenced, or such extended time as may be allowed under Clause 44 hereof.

43. 除按合同规定，工程整体完工前任何部分工程必须按时完工外，根据合同第 48 条的规定，整个工程必须在合同规定的期限内完成，该期限应从标书附录规定的开工期的最后一天算起，或以本合同第 44 条的规定而准许的延期时间为准。

44. Should the amount of extra or additional work of any kind or any cause of delay referred to in these Conditions, or exceptional adverse climatic conditions, or other special circumstances of any kind whatsoever which may occur, other than through a

default of the Contractor, be such as fairly to entitle the Contractor to an extension of time for the completion of the works, the Engineer shall determine the amount of such extension and shall notify the Employer and the contractor accordingly. Provided that the Engineer is not bound to take into account any extra or additional work or other special circumstances unless the Contractor has within _____ days after such work has been commenced, or such circumstances have arisen, or as soon thereafter as is practicable, submitted to the Engineer's Representative full and detailed particulars of any extension of time to which he may consider himself entitled in order that such submission may be investigated at the time.

44. 除承包人违约外，如因额外工程或工程量的增加或由此而涉及的任何延误原因、或特别恶劣的气候、或其他可能发生的任何特殊情况而使得承包人有权要求工程延期，工程师应决定延期时间，并由此通知业主和承包人。如承包人在此种工程开工后、或此种情况一发生、或按实情在其发生后的_____天内，没有向工程师代表提交详细说明他认为有权延期的报告而让此陈述当时便得到调查，工程师可不予考虑额外工程或工程量的增加或其他特殊情况。

45. Subject to any provision to the contrary contained in the Contract, none of the Permanent Works shall, save as hereinafter provided, be carried on during the night or on Sundays, if locally recognized as days of rest, or their locally recognized equivalent without the permission in writing of the Engineer's Representative, except when the work is unavoidable or absolutely necessary for the saving of life or property or for the safety of the works, in which case the Contractor shall immediately advise the Engineer's Representative. Provided always that the provisions of this Clause shall not be applicable in the case of any work which it is customary to carry out by rotary or double shifts.

45. 除合同另有相反规定外，未经工程师代表的书面允许，任何永久性工程不得在夜间或星期天进行，只要这些时间在当地被认作是假日，同时也不得在其他被当地认作是假日的时间进行，但下列情况除外：因工程而不得已或为挽救生命、财产或维护工程安全而绝对必要，在这种情况下，承包商应立即通知工程师代表。本条的规定不适用于按传统需轮班或倒班进行的工作。

46. If for any reason, which does not entitle the Contractor to an extension of time, the rate of progress of the works or any section is at any time, in the opinion of the Engineer, too slow to ensure completion by the prescribed time or extended time for completion, the Engineer shall so notify the Contractor in writing and the Contractor shall thereupon take such steps as are necessary and the Engineer may approve to

expedite progress so as to complete the Works or such section by the prescribed time or extended time. The Contractor shall not be entitled to any additional payment for taking such steps. If, as a result of any notice given by the Engineer under this Clause, the Contractor shall seek the Engineer's permission to do any work at night or on Sundays, if locally recognized as days of rest, or their locally recognized equivalent, such permission shall not be unreasonably refused.

46. 不论是何原因使得工程或其任何部分的进度让工程师觉得太慢，因而无法确保在规定或延长期限内完工，只要该原因不至于让承包人有权要求延期，工程师必须书面通知承包人，接到通知后承包人应采取必要的措施，工程师可同意加快进度，以便能在规定或延长期限内完工或完成某部分工程。承包人无权因采取此种措施而要求额外付款。如因工程师按本条款所发出的通知的缘故，承包人需要求工程师准许在夜间或星期日，如果其在当地被当作假日，或在其他被当地当作是假日的时间工作，工程师不得无故不准许。

47.1 If the contractor shall fail to achieve completion of the Works within the time prescribed by Clause 43 hereof, then the Contractor shall pay to the Employer the sum stated in the Contract as liquidated damages for such default and not as a penalty for every day or part of a day which shall elapse between the time prescribed by Clause 43 hereof and the date of certified completion of the Works. The Employer may, without prejudice to any other method of recovery, deduct the amount of such damages from any monies in his hands, due or which may become due to the contractor. The payment or deduction of such damages shall not relieve the Contractor from his obligation to complete the works, or from any other of his obligations and liabilities under the Contract.

47.1 如承包人未在本合同第43条规定的期限内完成工程，承包人应因违约而向业主支付合同所规定之款额，此款额为定额赔偿金，而不是根据第43条规定的期限与实际完工时间之差按天数进行处罚。业主可在不损害其他收款措施的情况下，从他手中掌握的应支付或将支付给承包人的款项中扣除此种赔偿金。此种赔偿金的支付或扣除不得免除承包人完成工程的义务，或免除他的其他任何合同所规定的责任和义务。

47.2 If, before the completion of the whole of the Works any part or section of the Works has been certified by the Engineer as completed, pursuant to Clause 48 hereof, and occupied or used by the Employer, the liquidated damages for delay shall, for any period of delay after such certificate and in the absence of alternative provisions in the Contract be reduced in the proportion which the value of the part or section so certified bears to the value of the whole of the Works.

47.2 在工程全部完工前，如有某部分工程已由工程师按本合同第48条确认为完工，且已被业主占用，如在此后出现任何延误，而本合同又无另行规定，因延误本应处罚的定额赔偿金，则应按已完工的工程的价值所占全部工程价值的比例予以扣减。

47.3　If it is desired to provide in the Contract for the payment of a bonus in relation to completion of the Works or of any part or section thereof this shall be set out in Part II in the clause numbered 47.

47.3　如要在合同中规定有关工程全部或部分完工的奖励支付事项，此规定应列在第二章的第47条中。

48.1　When the whole of the works have been substantially completed and have satisfactorily passed any final test that may be prescribed by the Contract, the Contractor may give a notice to that effect to the Engineer or to the Engineer's Representative accompanied by an undertaking to finish any outstanding work during the Period of Maintenance. Such notice and undertaking shall be in writing and shall be deemed to be a request by the Works. The Engineer shall, within _____ days of the date of delivery of such notice either issue to the contractor, with a copy to the Employer, a Certificate of Completion stating the date on which, in his opinion, the Works were substantially completed in accordance with the Contract or give instructions in writing to the Contractor specifying all the work which, in the Engineer's opinion, requires to be done by the Contractor before the issue of such certificate. The Engineer shall also notify the Contractor of any defects in the Works affecting substantial completion that may appear after such instructions and before completion of the works specified therein. The Contractor shall be entitled to receive such Certificate of Completion within _____ days of completion to the satisfaction of the Engineer of the works so specified and making good any defects so notified.

48.1　当工程全部完工且满意地通过了所有合同规定的最后检验后，承包人可向工程师或工程师代表发出有关通知，并承诺在维修期内将完成任何未完成的工作。此种通知和承诺必须做成书面形式，且应被视作是承包人要工程师发放工程竣工证书的要求。工程师应在此种通知送出后_____天内或向承包人发出工程竣工证书，并交业主一本副本，注明他认为工程已按合同规定圆满完工的日期，或向承包人发出书面指示，具体说明他认为在发放此种证书前承包人还需完成的一切工作。在发出此种指示后，工程师还应通知承包人关于此后以及在指示中所提及的工程完工前，可能出现且会影响工程实际完工的任何欠缺。在令工程师满意地完成所指明的工程并修正一切所指明的欠缺后，承包人有权在_____天之内收到竣工证书。

48.2　Similarly, in accordance with the procedure set out in sub-clause (1) of this Clause, the Contractor may request and the Engineer shall issue a Certificate of Completion in respect of a) any section of the Permanent Works in respect of which a separate time for completion is provided in the Contract and b) any substantial part of the Permanent Works which has been both completed to the satisfaction of the Engineer and occupied or used by the Employer.

48.2 同样，根据本条第1款规定的程序，就以下情况，承包人可要求且工程师应颁发竣工证书：合同中单独有完工期限规定的部分永久性工程竣工，以及永久性工程的任何实体部分已竣工，令工程师满意，且已被业主占有或使用。

48.3 If any part of the Permanent Works shall have been substantially completed and shall have satisfactorily passed any final test that may be prescribed by the Contract, the Engineer may issue a Certificate of Completion in respect of that part of the Permanent Works before completion of the whole of the Works and, upon the issue of such Certificate, the Contractor shall be deemed to have undertaken to complete any outstanding work in that part of the Works during the Period of Maintenance.

48.3 如部分永久性工程已实质性完工，且满意地通过了合同规定的最后检验，工程师可在全部工程竣工前颁发此部分工程的竣工证书，一旦颁发此种证书，承包人应被视作已承诺在维修期内完成此部分工程中一切尚未完成的工作。

48.4 Provided always that a Certificate of Completion given in respect of any section or part of the Permanent Works before completion of the whole shall not be deemed to certify completion of any ground or surfaces requiring reinstatement, unless such Certificate shall expressly so state.

48.4 在工程全部完工前就永久性工程的某一部分发放竣工证书，并不视为已经完成任何所需的地面或地表还原工作，证书中明文规定的除外。

MAINTENANCE AND DEFECTS
维修及欠缺

49.1 In these Conditions the expression "Period of Maintenance" shall mean the period of maintenance named in the Appendix to the Tender, calculated from the date of completion of the Works, certified by the Engineer in accordance with Clause 48 hereof, in the event of more than one certificate having been issued by the Engineer under the said Clause, from the respective dates so certified and in relation to the Period of Maintenance the expression "the Works" shall be construed accordingly.

49.1 本条款中"维修期"一词指标书附录所指的维修期,从工程师按本合同第 48 条规定确认的竣工日开始计算,如工程师根据上述条款发出了多份竣工证书,则分别从其确认的完工日开始计算,就维修期而言,"工程"一词须做相应解释。

49.2 To the intent that the Works shall, at or as soon as practicable after the expiration of the Period of Maintenance, be delivered to the Employer in the condition required by the Contract, fair wear and tear excepted, to the satisfaction of the Engineer, the Contractor shall finish the work, if any, outstanding at the date of completion, as certified under Clause 48 hereof, as soon as practicable after such date and shall execute all such work of repair, amendment, reconstruction, rectification and making good defects, imperfections, shrinkage or other faults as may be required of the Contractor in writing by the Engineer during the Period of Maintenance, or within _____ days after its expiration, as a result of an inspection made by or on behalf of the Engineer prior to its expiration.

49.2 为按合同规定条件在维修期满时或其后尽快将工程交给业主,除合理损耗外,为让工程师感到满意,承包人必须尽快完成本合同第 48 条中规定的在完工日尚未完成的工作(如果有),以及完成工程师在维护期间,或在维护期满后 _____ 天内,因工程师或工程师代表在维护期满前的检查而可能书面要求承包人完成的诸如修理、修正、再建、调整,以及修正欠缺、缺陷和其他毛病等一切工作。

49.3 All such work shall be carried out by the Contractor at his own expense if the necessity thereof shall, in the opinion of the Engineer, be due to the use of materials or workmanship not in accordance with the Contract, or to neglect or failure on the part of the Contractor to comply with any obligation, expressed or implied, on the Contractor's part under the Contract. If, in the opinion of the Engineer, such necessity shall be due to any other cause, the value of such work shall be ascertained and paid for as if it were additional work.

49.3 如工程师认为此种工作是因承包人使用的材料或工艺不符合合同的要求所致,或因承包人未履行其合同义务(明示或默示不论)所致,一切费用应由承包人承担。如工程师认为此种需要是其他原因所致,则应对此种工作进行估价,并按附加工程支付。

49.4 If the Contractor shall fail to do any such work as aforesaid required by the Engineer, the Employer shall be entitled to employ and pay other persons to carry out the same and if such work is work which, in the opinion of the Engineer, the Contractor was liable to do at his own expense under the Contract, then all expenses consequent thereon or incidental thereto shall be recoverable from the Contractor by the Employer, or may be deducted by the Employer from any monies due or which may become due to the Contractor.

49.4 如承包人未完成工程师要求的上述此种工作，业主有权雇用他人完成，如工程师认为，按合同规定此工作本应由承包人自费完成，业主应向承包人追偿由此发生的一切直接或间接费用，或从应付或可能应付给承包人的任何款项中予以扣除。

50. The Contractor shall, if required by the Engineer in writing, search under the directions of the Engineer for the cause of any defect, imperfection or fault appearing during the progress of the Works or in the Period of Maintenance. Unless such defect, imperfection or fault shall be one for which the Contractor is liable under the Contract, the cost of the work carried out by the Contractor in searching as aforesaid shall be borne by the Employer. If such defect, imperfection or fault shall be one for which the Contractor is liable as aforesaid, the cost of the work carried out in searching as aforesaid shall be borne by the Contractor and he shall in such case repair, rectify and make good such defect, imperfection or fault at his own expense in accordance with the provisions of Clause 49 hereof.

50. 应工程师的书面要求，承包人应根据工程师的指令，对施工过程或维修期内出现的任何欠缺、不足或缺陷进行检查。除非根据合同，此种欠缺、不足或缺陷应由承包人负责，否则上述检查工作的费用由业主承担。如此种欠缺、不足或缺陷如上所述属于承包人的责任，上述检查费用应由承包人负担，且他应按本合同第 49 条的规定，自费对其进行修理、修正和补救。

ALTERATIONS, ADDITIONS AND OMISSIONS
变更、增加和省略

51.1 The Engineer shall make any variation of the form, quality or quantity of the works or any part thereof that may, in his opinion, be necessary and for that purpose, or if for any other reason it shall, in his opinion be desirable, he shall have power to order the Contractor to do and the Contractor shall do any of the following: a) increase or decrease the quantity of any work included in the Contract, b) omit any such work, c) change the character or quality or kind of any such work, d) change the levels, lines, position and dimensions of any part of the Works, and e) execute additional work of any kind necessary for the completion of the Works and no such variation shall in any way vitiate or invalidate the Contract, but the value, if any, of all such variations shall be taken into account in ascertaining the amount of the Contract Price.

51.1 如工程师认为必要，他应对整个工程或任何部分的形式、质量或工作量做相应变更，且他有权因其他任何理由，命令承包人做出且承包人必须做出以下变更：增减合同所规定的工程量，省略任何部分工程，改变任何部分工程的特性或质量或类别，改变任何部分工程的平面、型线、位置和面积，增加竣工所需的任何额外工程，但此种变更不得以任何方式使合同失效，此种变更所需的一切费用（如果有）应在确定合同价格时予以考虑。

51.2 No such variations shall be made by the Contractor without an order in writing of the Engineer. Provided that no order in writing shall be required for increase or decrease in the quantity of any work where such increase or decrease is not the result of an order given under this Clause, but is the result of the quantities exceeding or being less than those stated in the Bill of Quantities. Provided also that if for any reason the Engineer shall consider it desirable to give any such order verbally, the Contractor shall comply with such order and any confirmation in writing of such verbal order given by the Engineer, whether before or after the carrying out of the order, shall be deemed to be an order in writing within the meaning of this Clause. Provided further that if the Contractor shall within _____ days confirm in writing to the Engineer and such confirmation shall not be contradicted in writing within _____ days by the Engineer, it shall be deemed to be an order in writing by the Engineer.

51.2 如无工程师的书面命令，承包人不得做任何此种变更。然而，如工程量的增减不属本条规定的应做命令的范畴，而是因超过或不足建筑工程清单的规定所致，则不要求有书面命令。另外，如工程师因任何原因认为有必要口头做此种命令，承包人必须服从，且工程师对该口头命令的确认书，不论在命令执行前或后给予，都应被视作是本条款所指的书面命令。此外，如承包人在_____天内书面向工程师要求确认，而工程师在_____天内未用书面驳回此确认，其应被视为是工程师所做的书面命令。

52.1 All extra or additional work done or work omitted by order of the Engineer shall be valued at the rates and prices set out in the Contract if, in the opinion of the Engineer, the same shall be applicable. If the Contract does not contain any rates or prices applicable to the extra or additional work, then suitable rates or prices shall be agreed upon between the Engineer and the Contractor. In the event of disagreement the Engineer shall fix such rates or prices as shall, in his opinion, be reasonable and proper.

52.1 凡由工程师命令增减的工程，如工程师认为可行，均应按合同价格予以估价。如合同所列的价格不适用所增减的工程，应由工程师和承包人协商出合适的价格。如不能就此达成协议，应由工程师决定一个他认为合理而恰当的价格。

52.2　Provided that if the nature or amount of any omission or addition relative to the nature or amount of the whole of the Works or to any part thereof shall be such that, in the opinion of the Engineer, the rate or price contained in the Contract for any item of the Works is, by reason of such omission or addition, rendered unreasonable or inapplicable, then a suitable rate or price shall be agreed upon between the Engineer and the Contractor. In the event of disagreement the Engineer shall fix such other rate or price as shall, in his opinion, be reasonable and proper having regard to the circumstances.

Provided also that no increase or decrease under sub-clause (1) of this Clause or variation of rate or price under sub-clause (2) of this Clause shall be made unless, as soon after the date of the order as is practicable and, in the case of extra or additional work, before the commencement of the work or as soon thereafter as is practicable, notice shall have been given in writing:

a) by the Contractor to the Engineer of his intention to claim extra payment or a varied rate or price, or

b) by the Engineer to the Contractor of his intention to vary a rate or price.

52.2　如工程师认为，涉及整个工程或部分工程的质或量的增减，使得本合同中所列的任何工程项目的价格因此增减而不再合理或不切实际，工程师和承包人则应协商一个合适的价格。如不能就此达成协议，应由工程师决定一个他认为合理而恰当地符合实际情况的价格。

除非在命令做出之后，如是额外或增加工程，则在工程动工前或之后尽快以书面形式：由承包人通知工程师他要求额外支付或更改价格，或由工程师通知承包人他打算改变价格。

否则不得按本条第1款增加或减少工程量，或按本条第2款变动价格。

52.3　If, on certified completion of the whole of the Works it shall be found that a reduction or increase greater than ＿＿＿＿ per cent of the sum named in the Letter of Acceptance, excluding all fixed sums, provisional sums and allowance for day works, if any, results from a) the aggregate effect of all Variation Orders, and b) all adjustments upon measurement of the estimated quantities set out in the Bill of Quantities, excluding all provisional sums, dayworks and adjustments of price made under Clause 70 (1) hereof, but not from any other cause, the amount of the Contract Price shall be adjusted by such sum as may be agreed between the Contractor and the Engineer or, failing agreement, fixed by the Engineer having regard to all material and relevant factors, including the Contractor's Site and general overhead costs of the Contract.

52.3 竣工验证时，如发现增减工程的费用超过验收证书所注明数额的_____%，除去所有的规定费用、临时费用和零工补贴外，如有任何是因：变动命令累计所致，以及因修正建筑工程清单所列的估算工程量所致，所有临时费用、零工补贴和本合同第 70 条第 1 款规定的价格调整除外，若非其他原因，合同价格应由承包人和工程师协商予以调整，或如协商不果，则由工程师在考虑所有材料及相关的包括承包人的场地及合同的一般间接费用在内的因素的基础上予以决定。

52.4 The Engineer may, if, in his opinion it is necessary or desirable, order in writing that any additional or substituted work shall be executed on a day-work basis. The Contractor shall then be paid for such work under the conditions set out in the day-work Schedule included in the Contract and at the rates and prices affixed thereto by him in his Tender.

The Contractor shall furnish to the Engineer such receipts or other vouchers as may be necessary to prove the amounts paid and, before ordering materials, shall submit to the Engineer quotations for the same for his approval.

In respect of all work executed on a day-work basis, the Contractor shall, during the continuance of such work, deliver each day to the Engineer's Representative an exact list in duplicate of the names, occupation and time of all workmen employed on such work and a statement, also in duplicate, showing the description and quantity of all materials and plant used thereon or therefor (other than plant which is included in the percentage addition in accordance with the Schedule hereinbefore referred to). One copy of each list and statement will, if correct, or when agreed, be signed by the Engineer's Representative and returned to the Contractor.

At the end of each month the Contractor shall deliver to the Engineer's Representative a priced statement of the labor, material and plant, except as aforesaid, used and the Contractor shall not be entitled to any payment unless such lists and statements have been fully and punctually rendered. Provided always that if the Engineer shall consider that for any reason the sending of such lists or statements by the Contractor, in accordance with the foregoing provision, was impracticable he shall nevertheless be entitled to authorize payment for such work, either as day-work, on being satisfied as to the time employed and plant and materials used on such work or at such value therefore as shall, in his opinion, be fair and reasonable.

52.4 如工程师认为有必要，可书面命令将增加或替换的工程按零工计价。此种工程承包人所得报酬应根据合同中的零工细目表的条件，按其标书中所附的价格予以支付。

承包人应向工程师提供必要的收据或其他凭证以证实其支出，并应在订购材料之前，向工程师提交报价单以求批准。

所有按零工计价的工程，在工程进行期间，承包人必须每天向工程师代表提供详细清单，一式两份，注明此种工程雇用的所有零工的姓名、职务和工作时间，以及两份说明书，标明用于工程的材料和设备的规格、数量（按上述细目表纳入增加费用计算内的设备除外）。如清单和说明书正确无误，工程师代表将认同且在其中一份上签字，然后退还给承包人。

每月底，承包人应向工程师代表提交一份所使用的劳力、材料和设备（上述规定的除外）的价目清单，如不按时提供完整的价目清单，承包人将无权得到支付。如工程师有理由认为承包人无法按上述规定提交清单，他将有权决定把此种工作或作为零工，按工时及使用的材料和设备进行支付，或按他认为公平合理的价格予以支付。

52.5　The Contractor shall send to the Engineer's Representative once in every month an account giving particulars, as full and detailed as possible, of all claims for any additional payment to which the Contractor may consider himself entitled and of all extra or additional work ordered by the Engineer which he has executed during the preceding month.

No final or interim claim for payment for any such work or expense will be considered which has not been included in such particulars. Provided always that the Engineer shall be entitled to authorize payment to be made for any such work or expense, notwithstanding the Contractor's failure to comply with this condition, if the Contractor has, at the earliest practicable opportunity, notified the Engineer in writing that he intends to make a claim for such work.

52.5　承包人应每月向工程师代表提交一份报告，尽可能全面详尽地索要他所认为有权要求的一切额外费用，以及陈述上月他按工程师的命令所做的一切额外工作。

凡未包括在报告中的工作或费用不能在中期或最后给予支付。但如承包人在最初曾书面通知工程师他想就此种工作要求索赔，即使他未遵守上述规定，工程师也有权要求对此种工作或费用给予支付。

PLANT，TEMPORARY WORKS AND MATERIALS
设备、临建工程和材料

53.1　All Constructional Plant, Temporary Works and materials provided by the Contractor shall, when brought on to the Site, be deemed to be exclusively intended for the execution of the Works and the Contractor shall not remove the same or any part thereof, except for the purpose of moving it from one part of the Site to another, without the consent, in writing, of the Engineer.

53.1 承包人所提供的所有建筑设备、临建工程设施和材料抵达工地后，应视作完全用于施工，如没有工程师书面同意，承包人不得移动它们或其任何部分，将它们从工地一端移到另一端除外。

53.2 Upon completion of the Works the Contractor shall remove from the Site all the said Constructional Plant and Temporary Works remaining thereon and any unused materials provided by the Contractor.

53.2 工程完工后，承包人应从工地将所有上述建筑设备、临建工程设施及按合同所提供的材料的剩余部分清理走。

53.3 The Employer shall not at any time be liable for the loss of or damage to any of the said constructional Plant, Temporary works or materials save as mentioned in Clauses 20 and 65 hereof.

53.3 业主对上述建筑材料、临建工程及材料的损失或损坏概不负责，本合同第20和第65条规定的除外。

53.4 In respect of any Constructional Plant which the Contractor shall have imported for the purposes of the works, the Employer will assist the Contractor, where required, in procuring any necessary Government consent to the re-export of such Constructional Plant by the Contractor upon the removal thereof as aforesaid.

53.4 如承包人因工程所需而要进口任何建筑设备，业主必须应承包人要求协助其从政府处得到必要的许可，以便按上述规定清理现场时重新出口此种建筑设备。

53.5 The Employer will assist the Contractor, where required, in obtaining clearance through the Customs of Constructional Plant, materials and other things required for the works.

53.5 业主应在需要时协助承包人为工程所需的建筑设备、材料和其他物品结关。

53.6 Any other conditions affecting Constructional Plant, Temporary Works and materials, shall be set out in Part II in the Clause numbered 53 as may be necessary.

53.6 其他任何影响建筑设备、临建工程设施和材料的条件必要时应在第二章的第53条予以规定。

54. The operation of Clause 53 hereof shall not be deemed to imply any approval by the Engineer of the materials or other matters referred to therein nor shall it prevent the rejection of any such materials at any time by the Engineer.

54. 本合同第53条的执行不得视为默示工程师认可该条款所指的材料或其他事项，也不得妨碍工程师在任何时候拒绝使用任何此种材料。

MEASUREMENT
测定

55. The quantities set out in the Bill of Quantities are the estimated quantities of the work, but they are not to be taken as the actual and correct quantities of the Works to be executed by the Contractor in fulfillment of his obligations under the Contract.

55. 建筑工程清单所列的工程量为估计数量，不应视为是承包人履行合同义务所完成的实际工程量。

56. The Engineer shall, except as otherwise stated, ascertain and determine by measurement the value in terms of the Contract of work done in accordance with the Contract. He shall, when he requires any part or parts of the Works to be measured, give notice to the Contractor's authorized agent or representative, who shall forthwith attend or send a qualified agent to assist the Engineer or the Engineer's Representative in making such measurement, and shall furnish all particulars required by either of them. Should the Contractor not attend, or neglect or omit to send such agent, then the measurement made by the Engineer or approved by him shall be taken to be the correct measurement of the work. For the purpose of measuring such permanent work as is to be measured by records and drawings, the Engineer's Representative shall prepare records and drawings month by month of such work and the Contractor, as and when called upon to do so in writing, shall, within _____ days, attend to examine and agree such records and drawings with the Engineer's Representative and shall sign the same when so agreed. If the Contractor does not so attend to examine and agree such records and drawings, they shall be taken to be correct. If, after examination of such records and drawings, the Contractor does not agree the same or does not sign the same as agreed, they shall nevertheless be taken to be correct, unless the Contractor shall, within _____ days of such examination, lodge with the Engineer's Representative, for decision by the Engineer, notice in writing of the respects in which such records and drawings are claimed by him to be incorrect.

56. 如无另行规定，工程师应通过测定判断按合同所完成的工程的价值。如他要求对某部分或某些工程进行测定，他应通知承包人的授权代理人或代表，该代理人或代表应立即参加或派一名合格的代理人协助工程师或工程师代表进行此种测定，并提供工程师或其代表所要求的一切详细情况。倘若承包人未参加或忽略或忘记派代理人协助，工程师所做的测定或他认可的测定，应被视为是对工程所做的正确测定。如永久性工程是靠记录和图纸进行测定，工程师代表应按月准备记录和图纸，如应书面要求，承包人

应当在 _____ 天内与工程师代表一起检查并认可此种记录和图纸，一经认可，承包人应在记录和图纸上签字。如果承包人不参与检查，记录和图纸应被视为正确无误。如检查完此种记录和图纸后，承包人不予认可或虽认可但不签字，该记录和图纸仍应被视为正确无误，除非承包人在此种检查后 _____ 天内书面告知工程师代表，他认为记录和图纸中哪些不正确，并要求工程师予以裁决。

57. The Works shall be measured net, notwithstanding any ganeral or local custom, except where otherwise specifically described or prescribed in the Contract.

57. 工程测定应采用净值，不管一般习惯或当地惯例如何，本合同另有明文规定的除外。

PROVISIONAL SUMS
暂列款

58.1 "Provisional Sum" means a sum included in the Contract and so designated in the Bill of Quantities for the execution of work or the supply of goods, materials, or services, or for contingencies, which sum may be used, in whole or in part, or not at all, at the direction and discretion of the Engineer. The Contract Price shall include only such amounts in respect of the work, supply or services to which such Provisional Sums relate as the Engineer shall approve or determine in accordance with this Clause.

58.1 "暂列款"是指合同中规定且列在建筑工程清单上的，按工程师的指令及自由处置的一笔用于施工，提供货物、材料或服务或用于意外事故的金额，可全部或部分使用，或根本不用。凡涉及与暂列款有关的工程、供应或服务的款项，只有经工程师按本合同规定同意或决定使用的方能列入合同价格。

58.2 In respect of every Provisional Sum the Engineer shall have power to order：

a) work to be executed, including goods, materials or services to be supplied by the Contractor. The Contract Price shall include the value of such work executed or such goods, materials or services supplied determined in accordance with Clause 52 hereof.

b) work to be executed or goods, materials or services to be supplied by a nominated Sub-Contractor as hereinafter defined. The sum to be paid to the Contractor therefor shall be determined and paid in accordance with Clause 59 (4) hereof.

c) goods and materials to be purchased by the Contractor. The sum to be paid to the Contractor therefor shall be determined and paid in accordance with Clause 59 (4) hereof.

58.2 就每一笔备用款，工程师有权命令用于：
   a) 施工，包括应由承包人提供的货物、材料或服务。合同价格应包括按合同第 52 条规定用于此种施工或提供的货物、材料或服务的费用。
   b) 由下文所规定的指定转包人的施工，提供的货物、材料或服务。由此付给承包人的款项应按本合同第 59 条第 4 款的规定确定并支付。
   c) 承包人所购买的货物和材料。由此支付给承包人的款项应按本合同第 59 条第 4 款确定并支付。

58.3 The Contractor shall, when required by the Engineer, produce all quotations, invoices, vouchers and accounts or receipts in connection with expenditure in respect of Provisional Sums.

58.3 凡工程师要求，承包人应提供与暂列款有关的开支的所有报价单、发票、凭证、账目或收据。

NOMINATED SUB-CONTRACTORS
指定分包人

59.1 All specialists, merchants, tradesmen and others executing any work or supplying any goods, materials or services for which Provisional Sums are included in the Contract, who may have been or be nominated or selected or approved by the Employer or the Engineer, and all persons to whom by virtue of the provisions of the Contract the Contractor is required to sub-let any work shall, in the execution of such work or the supply of such goods, materials or services, be deemed to be sub-contractors employed by the Contractor and are referred to in this Contract as "nominated Sub-Contractors".

59.1 所有在合同暂列款项下施工或提供任何货物、材料或服务的已由或将由业主或工程师指定或挑选的专家、商人、手工工人和其他人，以及所有根据合同规定，得到承包人任何分包工程且由此施工，提供货物、材料或服务的人，均应被视为是由承包人雇用的转包人，在本合同中称为"指定分包人"。

59.2 The Contractor shall not be required by the Employer or the Engineer or be deemed to be under any obligation to employ any nominated Sub-Contractor against whom the Contractor may raise reasonable objection, or who shall decline to enter into a sub-contract with the Contractor containing provisions:
   a) that in respect of the work, goods, materials or services the subject of the Sub-

contract, the nominated Sub-Contractor will undertake towards the Contractor the like obligations and liabilities as are imposed on the Contractor towards the Employer by the terms of the Contract and will save harmless and indemnify the Contractor from and against the same and from all claims, proceedings, damages, costs, charges and expenses whatsoever arising out of or in connection therewith, or arising out of or in connection with any failure to perform such obligations or to fulfil such liabilities, and

b) that the nominated Sub-Contractor will save harmless and indemnify the Contractor from and against any negligence by the nominated Sub-Contractor, his agents, workmen and servants and from and against any misuse by him or them of any Constructional Plant or Temporary Works provided by the Contractor for the purposes of the Contract and from all claims as aforesaid.

59.2 承包人不得因业主或工程师的要求，或被认为有任何义务而雇用其有理由反对的，或不与其签署含有下列规定的合同的指定分包人：

a) 就工程、货物、材料或服务，转包合同当事人，即指定分包人将对承包人承担本合同所规定的承包人应对业主承担的同样的义务和责任，并不让承包人再履行此义务和责任，且不让其受因履行此种义务和责任而产生或与之有关的，或因不履行此种义务或责任而产生或与之有关的一切索赔、诉讼、损害赔偿金、诉讼费、开支和费用的损害，以及

b) 指定分包人应使承包人不因指定分包人、其代理人、工人和雇员的过失而受任何损害，不因分包人、其代理人、工人和雇员不当使用承包人按合同所提供的工程设备或临建工程而受损害，以及不因上述所有的索赔而受损害。

59.3 If in connection with any Provisional Sum the services to be provided include any matter of design or specification of any part of the Permanent Works or of any equipment or plant to be incorporated therein, such requirement shall be expressly stated in the Contract and shall be included in any nominated Sub-Contract. The nominated Sub-Contract shall specify that the nominated Sub-Contractor providing such services will save harmless and indemnify the Contractor from and against the same and from all claims, proceedings, damages, costs, charges and expenses whatsoever arising out of or in connection with any failure to perform such obligations or to fulfil such liabilities.

59.3 如暂列款项下所提供的服务包括任何设计事项或对任何一部分永久性工程的详细规定，或说明工程所使用的任何设备的规格，此种要求应在合同中明文规定，且写进指定分包合同。指定分包合同应明确规定，提供此种服务的指定分包人应不让承包

人承担此种服务，且不让承包人因未履行此种义务或因履行此种义务而引起的或有关的任何索赔、诉讼、损害赔偿金、诉讼费、开支和费用而受损害。

59.4 For all work executed or goods, materials, or services supplied by any nominated Sub-Contractor, there shall be included in the Contract Price:

  a) the actual price paid or due to be paid by the Contractor, on the direction of the Engineer, and in accordance with the Sub-Contract;

  b) the sum, if any, entered in the Bill of Quantities for labor supplied by the Contractor in connection therewith, or if ordered by the Engineer pursuant to Clause 58(2)(b) hereof, as may be determined in accordance with Clause 52 hereof;

  c) in respect of all other charges and profit, a sum being a percentage rate of the actual price paid or due to be paid calculated, where provision has been made in the Bill of Quantities for a rate to be set against the relevant Provisional Sum, at the rate inserted by the Contractor against that item or, where no such provision has been made, at the rate inserted by the Contractor in the Appendix to the Tender and repeated where provision for such is made in a special item provided in the Bill of Quantities for such purpose.

59.4 就指定分包人所进行的所有施工，提供的货物、材料或服务而言，应纳入合同价格中的有：

  a) 按工程师的指令及根据分包合同，已实际由或应由承包人支付的款项；

  b) 涉及建筑工程清单中所列的（如果有）由承包人提供劳务的，或可能由工程师按本合同第58条第2款b）所命令的，且可根据本合同第52条确定的款项；

  c) 就所有其他费用和收益而言，应采用一百分比率乘以实际支付或即将支付的款额所得出款项计算，如建筑工程清单已有对有关暂定款制订了一个比率的规定，则应按承包人就该项目制订的比率予以计算，如无此种规定，则应按承包人在标书附录所列的，且建筑工程清单已就该具体项目做出认可的比率计算。

59.5 Before issuing, under Clause 60 hereof, any certificate, which includes any payment in respect of work done or goods, materials or services supplied by any nominated Sub-Contractor, the Engineer shall be entitled to demand from the Contractor reasonable proof that all payments, less retentions, included in previous certificates in respect of the work or goods, materials or services of such nominated Sub-Contractor have been paid or discharged by the Contractor, in default whereof unless the Contractor shall:

a) inform the Engineer in writing that he has reasonable cause for withholding or refusing to make such payments and

b) produce to the Engineer reasonable proof that he has so informed such nominated Sub-Contractor in writing, the Employer shall be entitled to pay to such nominated Sub-Contractor direct, upon the certificate of the Engineer, all payments, less retention, provided for in the Sub-Contract, which the Contractor has failed to make to such nominated Sub-Contractor and to deduct by way of set-off the amount so paid by the Employer from any sums due or which may become due from the Employer to the Contractor.

Provided always that, where the Engineer has certified and the Employer has paid direct as aforesaid, the Engineer shall, in issuing any further certificate in favor of the Contractor, deduct from the amount thereof the amount so paid, direct as aforesaid, but shall not withhold or delay the issue of the certificate itself when due to be issued under the terms of the Contract.

59.5 在按本合同第60条出具任何证书前，若其包括有关任何指定分包人所完成的工程，提供的货物、材料或服务的任何付款，工程师有权要求承包人提供充足证据，证明包括在先前证书中的有关该指定分包人的工程或货物、材料或服务的付款，已全部由承包人支付或清偿，否则将视为拖欠，除非承包人：

书面通知工程师，他有充足理由不予或拒绝此种付款支付，且向工程师提供充足证据，证明他已就此书面通知该指定分包人，业主有权在工程师出具证书后，向该指定分包人直接支付按规定应由承包人向该指定分包人支付，然而却没有支付的所有款项，留置款除外，然后从业主应支付或即将支付给承包人的任何款项中将此笔付款抵消。

如工程师已出具证书，且业主已直接做出上述支付，工程师在向承包人出具任何其他证书，应把业主直接支付的上述款项扣出，但工程师不得拒绝或延误出具按本合同规定应出具的证书本身。

59.6 In the event of a nominated Sub-Contractor, as hereinbefore defined, having undertaken towards the Contractor in respect of the work executed, or the goods, materials or services supplied by such nominated Sub-Contractor, any continuing obligation extending for a period exceeding that of the Period of Maintenance under the Contract, the Contractor shall at any time after the expiration Period of Maintenance, assign to the Employer, at the Employer's request and cost, the benefit of such obligation for the unexpired duration thereof.

59.6 如上所规定，如果指定分包人在施工，或提供货物、材料、或服务中对承包人承担的义务期限超过本合同所规定的维修期，承包人应在维修期满后，随时把此种义务在期满前所产生的利益应业主的要求转让给业主，费用由业主承担。

CERTIFICATES AND PAYMENT
证书和付款

60.1　Unless otherwise provided, payment shall be made at monthly intervals in accordance with the conditions set out in Part II in the Clause numbered 60.

60.1　除另有规定外，付款应按第二章第 60 条的规定，分月进行。

60.2　Where advances are to be made by the Employer to the Contractor in respect of Constructional Plant and materials, the conditions of payment and repayment shall be as set out in Part II in the Clause numbered 60.

60.2　如业主向承包人预付有关建筑设备和材料的款项，付款和还款的条件应在第二章第 60 条中予以规定。

60.3　If the execution of the Works shall necessitate the importation of materials, plant or equipment from a country other than that in which the Works are being executed, or if the Works or any part thereof are to be executed by labor imported from any other such country, or if any other circumstances shall render it necessary or desirable, a proportion of the payments to be made under the Contract shall be made in the appropriate foreign currencies and in accordance with the provisions of Clause 72 hereof. The conditions under which such payments are to be made shall be as set out in Part II in the Clause numbered 60.

60.3　如因施工需从施工所在地国的他国进口材料、设备，或因工程或任何部分工程需由此种他国输入的劳务完成，或因任何其他情况必需或必要，合同项下的某部分付款需按本合同第 72 条的规定用有关外汇进行支付，支付条件应按合同第二章第 60 条的规定处理。

61.　No certificate other than the Maintenance Certificate referred to in Clause 62 hereof shall be deemed to constitute approval of the Works.

61.　除本合同第 62 条所述的维修证书外，任何证书不得视为是对工程的认可。

62.1　The Contract shall not be considered as completed until a Maintenance Certificate shall have been signed by the Engineer and delivered to the Employer stating that the Works have been completed and maintained to his satisfaction. The Maintenance Certificate shall be given by the Engineer within _____ days after the expiration of the Period of Maintenance, or, if different periods of maintenance shall become applicable to different sections or parts of the Works, the expiration of the latest such period, or as soon thereafter as any works ordered during such period, pursuant to Clause 49 and 50 hereof, shall have been completed to the satisfaction of the Engineer and full effect shall be given to this Clause, notwithstanding any previous entry on the Works or the taking possession, working or using thereof or any part thereof by the Employer. Provided always that the issue of the Maintenance Certificate shall not be a condition precedent to payment to the Contractor of the second portion of the retention money in accordance with the conditions set out in Part Ⅱ in the Clause numbered 60.

62.1 只有工程师签发了维修证书且交给业主，说明工程已完工，维修情况令他满意，才能视合同已经完全履行。工程师应在维修期满后，或如工程的不同部分适用不同的维修期，在最后一个维修期满后_____天内签发维修证书，或在所有在此维修期中，按本合同第49和第50条规定，命令进行的工程完成且令工程师满意后立即签发，本条款必须全面执行，不管先前业主是否已进入、占有或使用工程或工程的任何部分。维修证书的签发不得作为按第二章第60条的规定向承包人支付第二笔留置款的前提条件。

62.2 The Employer shall not be liable to the Contractor for any matter or thing arising out of or in connection with the Contract or the execution of the Works, unless the Contractor shall have made a claim in writing in respect thereof before the giving of the Maintenance Certificate under this Clause.

62.2 业主不得因本合同或施工所产生或与之有关的任何事项对承包人负责，除非承包人在签发本条款规定的维修证书前，就此种事项已提出书面索赔。

62.3 Notwithstanding the issue of the Maintenance Certificate the Contractor and, subject to subclause (2) of this Clause, the Employer shall remain liable for the fulfillment of any obligation incurred under the provisions of the Contract prior to the issue of the Maintenance Certificate which remains unperformed at the time such Certificate is issued and, for the purposes of determining the nature and extent of any such obligation, the Contract shall be deemed to remain in force between the parties hereto.

62.3 尽管签发了维修证书，承包人以及业主（但应符合本条第2款的规定）仍应对发生在证书签发前而在证书签发时尚未履行的合同义务负责，为决定此种义务的性质及范畴，应本合同继续对合同双方有效。

REMEDIES AND POWERS
补救和权力

63.1　If the Contractor shall become bankrupt, or have a receiving order made against him, or shall present his petition in bankruptcy, or shall make an arrangement with or assignment in favor of his creditors, or shall agree to carry out the Contract under a committee of inspection of his creditors or, being a corporation, shall go into liquidation (other than a voluntary liquidation for the purposes of amalgamation or reconstruction), or if the Contractor shall assign the Contract, without the consent in writing of the Employer first obtained, or shall have an execution levied on his goods, or if the Engineer shall certify in writing to the Employer that in his opinion the Contractor:

a) has abandoned the Contract, or

b) without reasonable excuse has failed to commence the Works or has suspended the progress of the Works for _____ days after receiving from the Engineer written notice to proceed, or

c) has failed to remove materials from the Site or to pull down and replace work for _____ days after receiving from the Engineer written notice that the said materials or work had been condemned and rejected by the Engineer under these conditions, or

d) despite previous warnings by the Engineer, in writing, is not executing the Works in accordance with the Contract, or is persistently or flagrantly neglecting to carry out his obligations under the Contract, or

e) has, to the detriment of good workmanship, or in defiance of the Engineer's instructions to the contrary, sub-let any part of the Contract,

then the Employer may, after giving _____ days' notice in writing to the Contractor, enter upon the Site and the Works and expel the Contractor therefrom without thereby voiding the Contract, or releasing the Contractor from any of his obligations or liabilities under the Contract or affecting the rights and powers conferred on the Employer or the Engineer by the Contract, and may himself complete the Works or may employ any other contractor to complete the Works. The Employer or such other contractor may use for such completion so much of the Constructional Plant, Temporary Works and materials, which have been deemed to be reserved exclusively for the execution of the Works, under the provisions of the

Contract, as he or they may think proper, and the Employer may, at any time, sell any of the said Constructional Plant, Temporary Works and unused materials and apply the proceeds of sale in or towards the satisfaction of any sums due or which may become due to him from the Contractor under the Contract.

63.1 如果承包人破产，或被法院下达破产者产业管理接收令，或递交破产申请，或与债权人签订清偿协议或进行债权转让，或同意在债权人决议委任的破产监督委员会监督下执行本合同，或如其是一个公司，将被清算（不是因合并或重组而进行的自愿清算），或承包人事先未征得业主的书面同意而转让合同，或其货物被扣押，或工程师书面向业主证明，他认为承包人：

已经撤销合同，或没有正当理由但却不开工，或在收到工程师要求继续工程的书面通知后，停工_____天，或在收到工程师书面通知，说明某些材料不适用或某工程应予否决后_____天内未从工地搬走这些材料，或未拆毁此种工程重建，或不按合同施工，或长期或公然不履行其合同义务，尽管工程师已书面提出警告，或不顾工作质量，或无视工程师的指示，转包任何合同部分，此时，业主可在书面通知承包人_____天后进驻工地，并将承包人逐出，由此不会造成合同无效，或解除承包人的任何合同义务或责任，或影响合同赋予业主或工程师的权利和权力，业主可自己完成工程或另雇其他承包人完成工程。为完成工程，业主和其他承包人可充分使用其认为合适的，按本合同规定被视为是专门为施工而保留的建筑设备、临建工程和材料，业主可随时出售上述任何建筑设备、临建工程和未使用的材料，并将出售收入抵付按合同规定承包人应支付或可能支付给他的任何款项。

63.2 The Engineer shall, as soon as may be practicable after any such entry and expulsion by the Employer, fix and determine ex parte, or by or after reference to the parties, or after such investigation or inquiries as he may think fit to make or institute, and shall certify what amount, if any, had at the time of such entry and expulsion been reasonably earned by or would reasonably accrue to the Contractor in respect of work then actually done by him under the Contract and the value of any of the said unused or partially used materials, any Constructional Plant and any Temporary Works.

63.2 业主进驻工地并逐走承包人之后，工程师应尽快单方面，或通过了解双方当事人，或在做过他所认为恰当的调查或咨询之后做出安排或决定，并证明到此种进驻与被逐时为止，按事实上已完成的合同工程，承包人已获得或理应获得的款项（如果有），以及计算出上述任何未使用或只部分使用的材料、建筑设备和临建工程的价值。

63.3 If the Employer shall enter and expel the Contractor under this Clause, he shall not be liable to pay to the Contractor any money on account of the Contract until the expiration of the Period of Maintenance and thereafter until the costs of execution and maintenance, damages for delay in completion, if any, and all other expenses incurred by the Employer have been ascertained and the amount thereof certified by the Engineer. The Contractor shall then be entitled to receive only such sum or sums, if any, as the Engineer may certify would have been payable to him upon due completion by him after deducting the said amount. If such amount shall exceed the sum which would have been payable to the Contractor on due completion by him, then the Contractor shall, upon demand, pay to the Employer the amount of such excess and it shall be deemed a debt due by the Contractor to the Employer and shall be recoverable accordingly.

63.3 如果业主按本条规定进驻工地并逐走承包人，他应在维修期满，且所有施工和维修费、延误完工赔偿费（如果有）、和业主所发生的其他费用被核实，且经工程师证明后，方才负责对承包人按合同进行支付。此时，承包人有权得到的款额（如果有）仅为工程师证明，因他妥善完工而应得的款额减去上述款项之差。如上述费用超过承包人妥善完工应得的款额，经要求，承包人得向业主支付此超出部分款额，超出部分应视为承包人对业主的负债，且可照此收回。

64. If, by reason of any accident, or failure, or other event occurring to in or in connection with the Works, or any part thereof, either during the execution of the Works, or during the Period of Maintenance, any remedial or other work or repair shall, in the opinion of the Engineer or the Engineer's Representative, be urgently necessary for the safety of the Works and the Contractor is unable or unwilling at once to do such work or repair, the Employer may employ and pay other persons to carry out such work or repair as the Engineer or the Engineer's Representative may consider necessary. If the work or repair so done by the Employer is work which, in the opinion of the Engineer, the Contractor was liable to do at his own expense under the Contract, all expenses properly incurred by the Employer in so doing shall be recoverable from the Contractor by the Employer, or may be deducted by the Employer from any monies due or which may become due to the Contractor. Provided always that the Engineer or the Engineer's Representative, as the case may be, shall, as soon after the occurrence of any such emergency as may be reasonably practicable, notify the Contractor thereof in writing.

64. 如在施工或维修期间，因工程或部分工程发生或涉及任何意外、或事故、或其他事件，工程师或工程师代表认为需马上进行补救或其他工作或修理，以保证工程的安全，而承包人无法或不愿马上进行此种工作或修理，业主可雇用并支付他人进行工程师或工程师代表认为必要的此种工作或修理。如工程师认为，业主所进行的此种工作属于承包人按合同规定应自己付费完成的工作，业主由此所发生的一切费用应由业主向承包人收回，或从其应付给或可能应付给承包人的款项中扣除。在此种任何紧急事件发生后，工程师或工程师代表应根据情况，尽快书面告知承包人。

SPECIAL RISKS
特定风险

65. Notwithstanding anything in the Contract contained:

65. 本合同任何条款均服从以下规定：

65.1 The Contractor shall be under no liability whatsoever whether by way of indemnity or otherwise for or in respect of destruction of or damage to the Works, save to work condemned under the provisions of Clause 39 hereof prior to the occurrence of any special risk hereinafter mentioned, or to property whether of the Employer or third parties, or for or in respect of injury or loss of life which is the consequence of any special risk as hereinafter defined. The Employer shall indemnify and save harmless the Contractor against and from the same and against and from all claims, proceedings, damages, costs, charges and expenses whatsoever arising thereout or in connection therewith.

65.1 任何责任，不论属于赔偿或其他，也不论其是有关或涉及工程（以下提及的任何特定风险发生前，已按本合同第39条宣布为不适用的工程除外）、业主或第三方当事人的财产的毁坏或损坏，或有关或涉及人员伤亡，只要是因以下所规定的特定风险导致，承包人概不负责。业主必须保证承包人不受上述情况以及由此所引起的任何索赔、诉讼、损害赔偿金、诉讼费、开支和费用的损害。

65.2 If the Works or any materials on or near in transit to the Site, or any other property of the Contractor used or intended to be used for the purposes of the Works, shall sustain destruction or damage by reason of any of the said special risks the Contractor shall be entitled to payment for:

a) any permanent work and for any materials so destroyed or damaged, and, so far as may be required by the Engineer, or as may be necessary for the completion of the Works, on the basis of cost plus such profit as the Engineer may certify to be reasonable;

b) replacing or making good any such destruction or damage to the Works;

c) replacing or making good such materials or other property of the Contractor used or intended to be used for the purposes of the Works.

65.2 如工程或任何在工地、或从附近运往工地的材料，或承包人其他用于或旨在用于工地的任何财产因上述特定风险受到毁坏或损坏，承包人有权得到以下偿付：
因此而被毁坏或损坏的任何永久性工程或材料，以及，只要工程师要求，或工程完工所必需，按成本费加上工程师证明合理的利润为基准支付；恢复或补偿工程因此被毁坏或损坏的部分；恢复或补偿承包人用于或旨在于工程的此种材料或其他财产。

65.3 Destruction, damage, injury or loss of life caused by the explosion or impact whenever and wherever occurring of any mine, bomb, shell, grenade, or other projectile, missile, munitions, or explosive of war, shall be deemed to be a consequence of the said special risks.

65.3 因地雷、炸弹、炮弹、手榴弹、或其他射弹、导弹、弹药、或军用炸药的爆炸或冲击（不论时间和地点）所导致的毁坏、损害、受伤或死亡应被视为是上述特定风险的结果。

65.4 The Employer shall repay to the Contractor any increased cost of or incidental to the execution of the Works, other than such as may be attributable to the cost of reconstructing work condemned under the provisions of Clause 39 hereof, prior to the occurrence of any special risk, which is howsoever attributable to or consequent on or the result of or in any way whatsoever connected with the said special risks, subject however to the provisions in this Clause hereinafter contained in regard to outbreak of war, but the Contractor shall as soon as any such increase of cost shall come to his knowledge forthwith notify the Engineer thereof in writing.

65.4 除本条以下关于战争爆发的规定外，业主应偿还承包人因上述特定风险而发生的任何施工额外或附加费用，在任何特定风险发生前按合同第 39 条视为不适用的工程的重建费不在此列，承包人一旦知道此种费用增加，必须立即书面通知工程师。

65.5 The special risks are war, hostilities (whether war be declared or not), invasion, act of foreign enemies, the nuclear and pressure-waves risk described in Clause 20 (2) hereof, or insofar as it relates to the country in which the Works are being or are to be executed or maintained, rebellion, revolution, insurrection, military or usurped power, civil war, or, unless solely restricted to the employees of the Contractor or of his Sub-Contractors and arising from the conduct of the Works, riot, commotion or disorder.

65.5 特定风险指战争、战争状态（不论宣战与否）、侵略、外国敌对行为、本合同第 20 条第 2 款规定的核危险和声波风险、或与正在或即将施工或维修的工程所在地国有关的反叛、革命、暴动、军事政变、篡权、内战，或（只是由承包人或其转包人的雇员和因施工引起的除外）骚乱、动乱或混乱。

65.6　If, during the currency of the Contract, there shall be an outbreak of war, whether war is declared or not, in any part of the world which, whether financially or otherwise, materially affects the execution of the Works, the Contractor shall, unless and until the Contract is terminated under the provisions of this Clause, continue to use his best endeavors to complete the execution of the Works. Provided always that the Employer shall be entitled at any time after such outbreak of war to terminate the Contract by giving written notice to the Contractor and, upon such notice being given, this Contract shall, except as to the rights of the parties under this Clause and to the operation of Clause 67 hereof, terminate, but without prejudice to the rights of either party in respect of any antecedent breach thereof.

65.6　在合同执行时，如世界任何地方爆发战争，无论宣战与否，是否在财政或其他方面对工程造成了重大影响，除非并在本合同按本条款规定终止前，承包人应继续尽力完成工程。业主在战争爆发后随时有权书面通知承包人终止合同，一旦此种通知发出，除本条所规定的双方权利和第67条的规定外，合同应当终止，但不得损害任何一方与以前任何违约有关的一切权利。

65.7　If the Contract shall be terminated under the provisions of the last preceding sub-clause, the Contractor shall, with all reasonable dispatch, remove from the Site all Constructional Plant and shall give similar facilities to his Sub-Contractors to do so.

65.7　如合同按上一款规定终止，承包人应迅速从工地移走所有建筑设备，并向其转包人提供同样的转移设施。

65.8　If the Contract shall be terminated as aforesaid, the Contractor shall be paid by the Employer, insofar as such amounts or items shall not have already been covered by payments on account made to the Contractor, for all work executed prior to the date of termination at the rates and prices provided in the Contract and in addition:

　　a) the amounts payable in respect of any preliminary items, as far as the work or service comprised therein has been carried out or performed, and a proper proportion as certified by the Engineer of any such items, the work or service comprised in which has been partially carried out or performed;

　　b) the cost of materials or goods reasonably ordered for the Works which shall have been delivered to the Contractor or of which the Contractor is legally liable to accept delivery, such materials or goods becoming the property of the Employer upon such payments being made by him;

　　c) a sum to be certified by the Engineer, being the amount of any expenditure reasonably incurred by the Contractor in the expectation of completing the whole of the Works insofar as such expenditure shall not have been covered by the payments in this sub-clause before mentioned;

d) any additional sum payable under the provisions of sub-clauses (1), (2) and (4) of this Clause;

e) the reasonable cost of removal of Constructional Plant under sub-clause (7) of this Clause and, if required by the Contractor, return thereof the Contractor's main plant yard in his country of registration or to other destination, at no greater cost;

f) the reasonable cost of repatriation of all the Contractor's staff and workmen employed on or in connection with the Works at the time of such termination.

Provided always that against any payments due from the Employer under this sub-clause the Employer shall be entitled to be credited with any outstanding balances due from the Contractor for advances in respect of Constructional Plant and materials and any other sums which at the date of termination were recoverable by the Employer from the Contractor under the terms of the Contract.

65.8 如果合同如上终止，就承包人所得的暂付款项中尚未包括的在终止之日前已完成工作的款项，业主应按本合同所规定的价格支付给承包人，此外还应支付：

a) 所有有关临时单据应予支付的款项，只要单据所含的工作或服务已经完成，如此种单据所含的工作或服务只完成一部分，则只支付经工程师确认的应予支付的那一部分款项；

b) 因工程而订购的材料或货物的合理费用，包括已运交承包人的和承包人按法律有义务接收的材料或货物，业主向承包人付款后，这些材料或货物即成为业主的财产；

c) 经工程师确认的，承包人为完成整个工程而发生的合理费用，只要本款上述各种费用未将此费用包括在内；

d) 本条第1、第2和第4款所规定的追加付款；

e) 本条第7款所规定的搬迁建筑设备的合理费用，如果承包人要求，包括将此设备运回到承包人注册国的主要设备堆置场的费用，或运至其他地点的费用，但此费用不得超过运回注册国的费用；

f) 合同终止时，遣返承包人雇用在工程上或与之有关的职员和工人的合理费用。

对本款项下业主应予支付的款项，业主有权用承包人的应付款来抵冲，包括建筑设备及材料的预付款的未偿余额，以及其他任何在合同终止日按合同条款业主有权向承包人回收的款项。

FRUSTRATION
受挫失效

66. If a war, or other circumstances outside the control of both parties, arises after the Contract is made so that either party is prevented from fulfilling his contractual obligations, or under the law governing the Contract, the parties are released from further performance, then the sum payable by the Employer to the Contractor in respect of the work executed shall be the same as that which would have been payable under Clause 65 hereof if the Contract had been terminated under the provisions of Clause 65 hereof.

66. 如合同缔结后爆发战争,或出现双方无法控制的其他情况,使得任何一方无法履行其合同义务,或根据合同适用的法律,双方被解除了继续履约的义务,由业主对已完成的工程而支付承包人的款额应当等于按本合同第65条规定之金额,即如果合同根据第65条之规定终止时,应当支付的金额。

SETTLEMENT OF DISPUTES
争议的解决

67. If any dispute or difference of any kind whatsoever shall arise between the Employer and the Contractor or the Engineer and the Contractor in connection with, or arising out of, the Contract or the execution of the Works, whether during the progress of the Works or after their completion and whether before or after the termination, abandonment or breach of the Contract, it shall, in the first place, be referred to and settled by the Engineer who shall, within a period of _____ days after being requested by either party to do so, give written notice of his decision to the Employer and the Contractor. Subject to arbitration, as hereinafter provided, such decision in respect of every matter so referred shall be final and binding upon the employer and the Contractor and shall forthwith be given effect to by the Employer and by the Contractor, who shall proceed with the execution of the Works with all due diligence whether he or the Employer requires arbitration, as hereinafter provided, or not. If the Engineer has given written notice of his decision to the Employer and the Contractor and no claim to arbitration has been communicated to him by either the Employer or the Contractor within a period of _____ days from receipt of such notice, the said decision shall remain final and binding upon the Employer and the Contractor. If the Engineer shall fail to give notice of his decision, as aforesaid,

within a period of _____ days after being requested as aforesaid, or if either the Employer or the Contractor be dissatisfied with any such decision, then and in any such case either the Employer or the Contractor may within _____ days after receiving notice of such decision, or within _____ days after the expiration of the first-named period of _____ days, as the case may be, require that the matter or matters in dispute be referred to arbitration as hereinafter provided. All disputes or differences in respect of which the decision, if any, of the Engineer has not become final and binding as aforesaid shall be finally settled under the Rules of Conciliation and Arbitration of the International Chamber of Commerce by one or more arbitrators appointed under such Rules. The said arbitrator/s shall have full power to open up, revise and review any decision, opinion, direction, certificate or valuation of the Engineer. Neither party shall be limited in the proceedings before such arbitrator/s to the evidence or arguments put before the Engineer for the purpose of obtaining his said decision. No decision given by the Engineer in accordance with the foregoing provisions shall disqualify him from being called as a witness and giving evidence before the arbitrator/s on any matter whatsoever relevant to the dispute or difference referred to the arbitrator/s as aforesaid. The reference to arbitration may proceed notwithstanding that the Works shall not then be or be alleged to be complete, provided always that the obligations of the Employer, the Engineer and the Contractor shall not be altered by reason of the arbitration being conducted during the progress of the Works.

67. 业主与承包人或工程师与承包人因合同或工程施工而发生的或与之有关的任何争议或分歧，无论是在施工中还是在完工后，无论是在合同终止、撤销或违约前后，首先应当提交工程师解决，应任何一方要求后_____天内，工程师必须将其决定书面通知业主和承包人。除按以下规定提出仲裁外，此种有关上述所有事项的决定为最终决定，对业主和承包人均有约束力的，且立即对业主和承包人生效，承包人应继续施工并尽一切应有的注意，无论他或业主或承包人是否要按以下规定提请仲裁。在工程师将决定书面通知业主和承包人后，如_____天内未收到业主或承包人的仲裁申请，上述决定应作为最终决定，对业主和承包人均有约束力。如工程师未按上述规定应要求在_____天内通知其决定，或业主或承包人对此种决定不满意，在此后或此种情况下，业主或承包人可在收到决定通知_____天内，或在最先所说的_____天期满后_____天内，视情况而定，将争议事项按下述规定提交仲裁。所有争议或分歧，凡工程师的有关决定（如果有）如上所述不再是最终和有约束力的，均应由根据《国际商会调解和仲裁规则》所指定的一名或多名仲裁员按该规则规定予以最终解决。上述仲裁员（们）有权评论、修正和复审工程师的任何决定、意见、指示、证书或估价。在仲裁过程中，任何一方均可不受限制地向仲裁员（们）提供当时供工程师做出决定而采用的证据或论据。不论工程师按上述规定曾做过何种决定，均不得剥夺其被传唤作为证人，且就按上述规定提交仲裁的争议或分歧的任何有关事项向仲裁员（们）提供证据的资格。不论工程尚未或认为尚未完工，均可提请仲裁，但业主、工程师和承包人的义务不得因在施工进程中提起仲裁而改变。

NOTICE
通知

68.1 All certificates, notices or written orders to be given by the Employer or by the Engineer to the Contractor under the terms of the Contract shall be served by sending by post to or delivering the same to the Contractor's principal place of business, or such other address as the Contractor shall nominate for this purpose.

68.1 业主或工程师按合同条款给予承包人的所有证书、通知或书面命令均应邮寄或送交至承包人的主要营业地，或承包人指定的其他通讯地址。

68.2 All notices to be given to the Employer or to the Engineer under the terms of the Contract shall be served by sending by post or delivering the same to the respective addresses nominated for that purpose in Part II of these Conditions.

68.2 所有按本合同条款给予业主或工程师的通知均应邮寄或送交至第二章有关通知条款所指定的其各自的地址。

68.3 Either party may change a nominate address to another address in the country where the Works are being executed by prior written notice to the other party and the Engineer may do so by prior written notice to both parties.

68.3 任何一方均可把一指定地址改为在施工地国的另外一个地址，但事先应书面通知另一方，工程师在改变地址前需书面通知双方当事人。

DEFAULT OF EMPLOYER
业主违约

69.1 In the event of the Employer：
   a) failing to pay to the Contractor the amount due under any certificate of the Engineer within _____ days after the same shall have become due under the terms of the Contract，subject to any deduction that the Employer is entitled to make under the Contract，or
   b) interfering with or obstructing or refusing any required approval to the issue of

any such certificate, or

c) becoming bankrupt or, being a company, going into liquidation, other than for the purpose of a scheme of reconstruction or amalgamation, or

d) giving formal notice to the Contractor that for unforeseen reasons, due to economic dislocation, it is impossible for him to continue to meet his contractual obligations, the Contractor shall be entitled to terminate his employment under the Contract after giving _____ days prior written notice to the Employer, with a copy to the Engineer.

69.1 如果业主：
在工程师证书所认定的应付款项按合同条款到期_____天内未向承包人支付，业主按合同有权所做的扣除部分除外，或干涉或阻碍或拒绝同意颁发任何此种证书，或破产或如为公司，即将被清算，而此种清算不是为了重组或合并，或正式通知承包人，因不可预见的原因，鉴于经济混乱，其无法再履行其合同义务，承包人应有权在提前_____天书面通知业主，并将副本呈交工程师后，终止其合同雇佣关系。

69.2 Upon the expiry of the _____ days' notice referred to in sub-clause (1) of this Clause, the Contractor shall, notwithstanding the provisions of Clause 53 (1) hereof, with all reasonable dispatch, remove from the Site all Constructional Plant brought by him thereon.

69.2 本条第1款规定的_____天通知期限到期后，尽管本合同第53条第1款做有规定，承包人仍然应尽量迅速地从工地撤走他所带来的所有建筑设备。

69.3 In the event of such termination the Employer shall be under the same obligations to the Contractor in regard to payment as if the Contract had been terminated under the provisions of Clause 65 hereof, but, in addition to the payments specified in Clause 65 (8) hereof, the Employer shall pay to the Contractor the amount of any loss or damage to the Contractor arising out of or in connection with or by consequence of such termination.

69.3 如果合同如此终止，业主必须向承包人承担本合同按第65条规定终止时一样的支付义务，但除了第65条第8款所规定的款项外，业主还必须向承包人支付因合同由此终止而造成的或有关的或导致的灭失或损失费。

CHANGES IN COSTS AND LEGISLATION
费用与法规的变更

70.1 Adjustments to the Contract Price shall be made in respect of rise or fall in the costs of labor and/or materials or any other matters affecting the cost of the execution of the Works, as set out in Part II in the Clause numbered 70.

70.1 合同价格的调整应与劳工或/和材料成本的增减，或其他任何影响施工费用的事项相关，应按第二章第 70 条规定执行。

70.2 If, after the date _____ days prior to the latest date for submission of tenders for the Works there occur in the country in which the Works are being or are to be executed changes to any National or State Statute, Ordinance, Decree or other Law or any regulation or by-law of any local or other duly constituted authority, or the introduction of any such State Statute, Ordinance, Decree, Law, regulation or by-law which causes additional or reduced cost to the Contractor, other than under sub-clause (1) of this Clause, in the execution of the Works such additional or reduced cost shall be certified by the Engineer and shall be paid by or credited to the Employer and the Contract Price adjusted accordingly.

70.2 在提交工程标书最后期限前的_____天之后，如施工或即将施工工程所在地国的任何国家或州的法规、法令、政令或其他任何地方和其他权力机构的法律或条例或规章发生变化，或适用了使承包人的施工费用发生增减（非本条第 1 款所述的增减）的任何州的法规法令、政令、法律、条例或规章，此种费用增减应经工程师证明，且由业主支付或记入贷方，合同价格也应做相应调整。

CURRENCY AND RATES OF EXCHANGE
货币和汇率

71. If, after the date _____ days prior to the latest date for submission of tenders for the Works the Government or authorized agency of the Government of the country in which the Works are being or are to be executed imposes currency restrictions and/or transfer of currency restrictions in relation to the currency or currencies in which the Contract Price is to be paid, the Employer shall reimburse any loss or damage to the Contractor arising therefrom, without prejudice to the right of the Contractor to exercise any other rights or remedies to which he is entitled in such event.

71. 在提交标书最后期限前的_____天之后，如施工或即将施工工程所在地国的政府或政府授权机构对合同价格计价货币实行货币管制和/或货币兑换管制，业主应补偿由此对承包人造成的一切损失或损害，承包人在此种情况下有权行使其他任何权利或补救措施不得由此而受影响。

72.1 Where the Contract provides for payment in whole or in part to be made to the

Contractor in foreign currency or currencies, such payment shall not be subject to variations in the rate or rates of exchange between such specified foreign currency or currencies and the currency of the country in which the Works are to be executed.

72.1 如果合同规定，承包人的支付应全部或部分使用一种或多种外币，此种支付不得受特定外币与施工工程所在地国的货币间的汇率变化的影响。

72.2 Where the Employer shall have required the Tender to be expressed in a single currency but with payment to be made in more than one currency and the Contractor has stated the proportions or amounts of other currency or currencies in which he requires payment to be made, the rate or rates of exchange applicable for calculating the payment of such proportions or amounts shall be those prevailing, as determined by the Central Bank of the country in which the Works are to be executed, on the date _____ days prior to the latest date for the submission of tenders for the Works, as shall have been notified to the Contractor by the Employer prior to the submission of tenders or as provided for in the tender documents.

72.2 如业主要求标书计价只用一种货币，而支付却用多种货币，且承包人已说明他要求用其他货币支付的比例或数额，用于计算此种比例或数额支付的货币兑换率应为现行汇率，即由施工所在地国的中央银行在提交标书最后期限前的_____天当日所决定的汇率，就此，业主必须在标书提交前告知承包人，或在投标文件中予以写明。

72.3 Where the Contract provides for payment in more than one currency, the proportions or amounts to be paid in foreign currencies in respect of Provisional Sum items shall be determined in accordance with the principles set forth in sub-clause (1) and (2) of this Clause as and when these sums are utilized in whole or in part in accordance with the provisions of Clause 58 and 59 hereof.

72.3 如合同规定使用多种货币支付，凡涉及按本合同第58和第59条规定使用暂列款的项目，其用外汇支付的比例或数额应按本条第1和第2款制定的原则予以确定。

PART Ⅱ Conditions of Particular Application
第二章 专用条款

（主要涉及一些具体情况的应用，限于篇幅，不列入第二章内容。）

References（参考文献）

[1] 俞家欢. 土木工程专业英语［M］. 北京：清华大学出版社，2017.
[2] 钱永梅. 新编土木工程专业英语［M］. 北京：化学工业出版社，2014.